SERIOUS MANAGERS GUIDE TO AI NAVIGATION OF FEDERAL HEALTHCARE

Full How–to Manual for Successful AI Deployments
in the Complex World of Federal Healthcare

By
Albert Villegas

Cybersoft Publishing LLC
Fort Washington Md 20744
May Edition April 2025

Table of Contents

1   Introduction

The U.S. healthcare system has been marred by challenges over the years, the most significant being high healthcare costs and inequitable access to care across population groups. Many U.S. families and individuals feel burdened by the high healthcare costs, leading to the demand for consideration of this factor in leadership and policy decisions. According to Peter G. Peterson Foundation (2024), more than half of the adults in the U.S. find it difficult to afford healthcare costs, while one quarter of the adult population have found problems paying for healthcare within the past 12 months. Healthcare costs disproportionately impact young adults, low-income individuals, adults with chronic conditions, minority populations, and the uninsured, making it harder for them to afford care compared to others. The high cost of healthcare is associated with poor health-seeking tendencies among the disproportionately affected populations as people often put off needed care due to lack of funds (Peter G. Peterson Foundation, 2024). Notably, 61% of uninsured adults reported going without needed care within the last 12 months (Lopes et al., 2024). These issues constantly affect healthcare services, access and delivery.

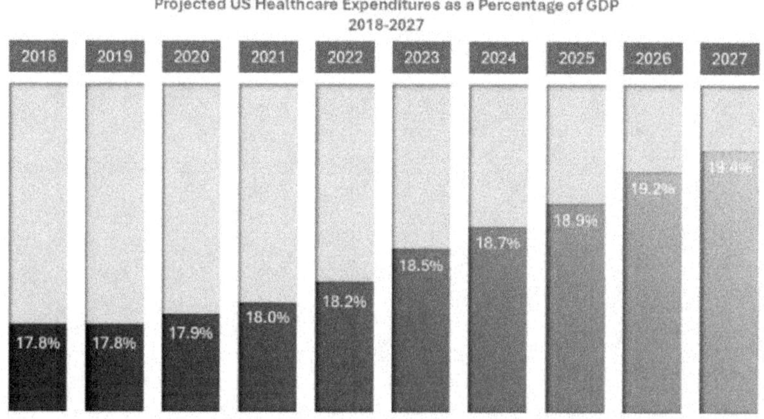

Figure 1 Healthcare Cost as a Percentage of GDP

Since its inception, Medicare has aimed at promoting healthcare access by making healthcare costs more affordable, a practice that should result in more equitable healthcare service delivery. Medicare goals have been centered on making healthcare more accessible to diverse populations, including the elderly, those with chronic conditions, the low-income communities, and other disproportionately affected populations (CMS, n.d.). Through its goals, Medicare has so far ensured improvements in healthcare access nationwide. The goals of Medicare foster movement toward healthcare equity by implementing various care principles. One of the strategies used recently to enhance access to care and equity through Medicare has been to focus on value-based care. Value-based care improves healthcare and patient experience by promoting provider collaboration, addressing patients holistically, and emphasizing both medical and non-medical needs. Simply implementing these practices makes it more likely that patients who might otherwise face difficulties accessing healthcare services for various reasons will seek and receive high-quality care. Because of the emphasis on value-based care, Medicare can provide more healthcare benefits and equitable billing for all patients.

Please note the following information, as published in 2022 by the Organization for Economic Co-operation (OECD)

- The U.S. spends more on healthcare as a share of the economy – nearly twice as much as the average OECD country – yet has the lowest life expectancy and the highest suicide rates among the 11 nations.

- The U.S has the highest chronic disease burden and an obesity rate that is two times higher than the OECD average

- Compared to peer nations, the U.S. has among the highest rates of hospitalizations from preventable causes and the highest rate of avoidable deaths.

Figures on Healthcare Spending

The U.S spent 4.1 TRILLION on health care in 2020

Hospital Care	1,270.1	30.8%
Other Personal healthcare	609.2	14.8%
Physician services	593.1	14.4%
Prescription drugs	348.4	8.4%
Net cost of health insurance	301.4	7.3%
Government public health activities	223.7	5.4%
Clinical services	216.3	5.2%
Nursing care facilities	196.8	4.8%
Investment	192.7	4.7%
Home health care	123.7	3.0%
Government administration	48.4	1.2%

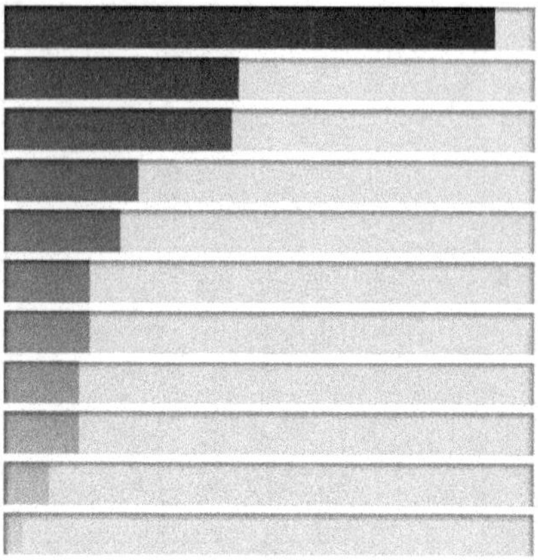

Figure 2 HHS on Healthcare Spending

2 Why AI in Government Healthcare Is Different

You're not managing AI deployment at a startup. You're not optimizing customer engagement for a consumer app. You're operating in an environment where mission comes first, where accountability is nonnegotiable, where every decision affects real lives—service members, veterans, beneficiaries, citizens.

This changes everything.

Federal healthcare AI isn't about moving fast and breaking things. It's about moving deliberately and fixing things—care delivery, operational efficiency, workforce capacity, mission readiness. It's about deploying technology that serves the public good while navigating constraints that would paralyze most private sector initiatives: procurement rules, budget cycles, interoperability mandates, oversight requirements, and a zero-tolerance environment for failure.

This chapter establishes why federal AI requires a fundamentally different playbook. It explores the unique constraints, opportunities, and responsibilities that define AI adoption in government healthcare. By the end, you'll understand why the approaches that work in Silicon Valley don't translate to federal operations—and what works instead.

2.1 The Mission-First Mandate

In the private sector, AI serves the bottom line. In federal healthcare, AI serves the mission. That mission varies by agency—readiness for the Defense Health Agency, lifetime care for the Department of Veterans Affairs, population health for the Centers for Medicare and Medicaid Services—but the principle remains constant: technology is a means to the mission, not an end in itself.

This mission-first orientation fundamentally shapes how you evaluate, procure, deploy, and govern AI systems.

2.2 Mission Defines Value

Private-sector AI projects justify themselves through revenue growth, cost reduction, or market-share expansion. Federal healthcare AI must demonstrate mission impact. What does that look like in practice?

For DHA, mission impact means operational readiness. An AI system that optimizes surgical scheduling isn't valuable because it reduces costs—it's valuable because it ensures combat-ready personnel receive timely care, maintains medical workforce readiness, and supports global deployments.

For VA, mission impact means honoring the commitment to veterans. An AI system that accelerates disability claims processing isn't about efficiency metrics—it's about reducing veteran wait times, improving access to earned benefits, and delivering on a national promise.

For CMS, mission impact means protecting beneficiaries and stewarding public resources. An AI system that detects fraud, waste, and abuse isn't about recovering dollars—it's about ensuring Medicare and Medicaid funds serve their intended purpose and protecting vulnerable populations from predatory practices.

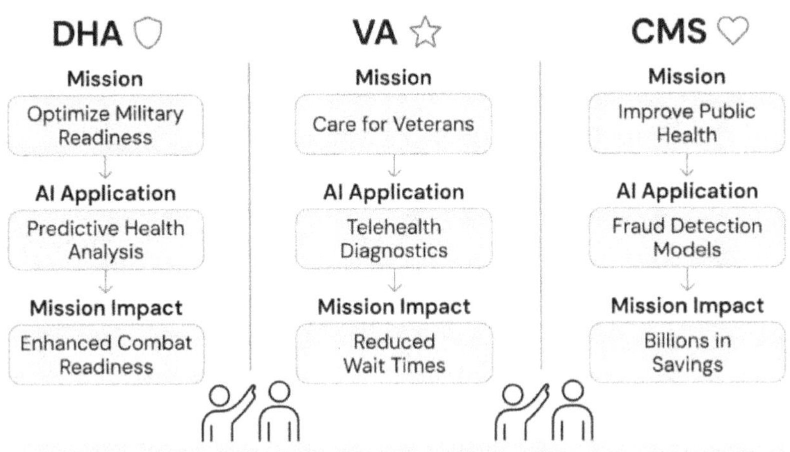

2.3   Mission Constraints Acceptable Risk

Mission primacy also defines your risk tolerance. In commercial AI, acceptable risk is calculated against potential returns. An 85 percent accurate recommendation engine might be acceptable if it drives sufficient engagement. In federal healthcare, risk is calculated against mission consequences.

Consider clinical decision support AI. An 85 percent accuracy rate means that 15 percent of recommendations in a commercial health app are wrong, which might prompt a customer to seek a second opinion. In a combat support hospital, this could delay critical treatment. In a VA facility serving elderly veterans with complex comorbidities, this could trigger inappropriate interventions.

The same logic applies to administrative systems. An AI chatbot that occasionally provides incorrect information might be acceptable for a retail customer service operation. For a veteran trying to understand their benefits, or a Medicare beneficiary navigating coverage options, incorrect information isn't an inconvenience—it's a potential crisis.

This doesn't mean federal healthcare can't use AI. It means federal healthcare AI must meet mission-grade reliability standards. You need higher accuracy thresholds, more robust validation, clearer fallback procedures, and stronger human oversight mechanisms.

2.4   Mission Demands Transparency

Federal operations occur in the public trust. Citizens, service members, veterans, and beneficiaries have the right to understand how decisions affecting them are made. This creates a transparency requirement that commercial AI rarely faces.

When an AI system recommends a course of treatment, schedules an appointment, or flags a claim for review, affected individuals and their advocates will ask: Why? How was this decision made? What data informed it? Can it be appealed?

Black-box AI systems—models that can't explain their reasoning— are incompatible with this requirement. You need explainable AI, audit trails, decision documentation, and clear escalation pathways. You need systems designed for scrutiny, not just performance.

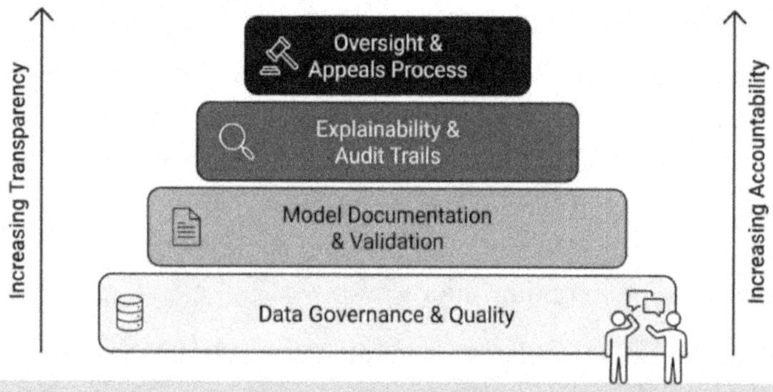

Public Trust and Accountability

Federal healthcare operates under a social contract. Service members trust that the military health system will maintain their readiness. Veterans trust that VA will honor service-connected commitments. Beneficiaries trust that Medicare and Medicaid will protect their health and welfare. This trust is the foundation of your operational legitimacy.

AI introduces new risks to that trust. Algorithmic bias, data breaches, unexplained decisions, and system failures can quickly erode confidence. In the private sector, companies weather these storms through marketing, rebranding, or pivoting. In federal healthcare, you don't have those options. Trust, once lost, is extraordinarily difficult to rebuild.

The Accountability Environment

Federal managers operate in an accountability environment that has no private sector equivalent. You're accountable to:

- Agency leadership and organizational chains of command
- Congressional oversight committees and appropriations processes
- Inspector General offices and audit functions
- Office of Management and Budget requirements
- Government Accountability Office reviews
- Media scrutiny and public records requests
- Advocacy organizations and stakeholder groups
- Legal and regulatory compliance frameworks

Every AI system you deploy will be examined through these lenses. Not occasionally—continuously. This creates a documentation burden, an approval hierarchy, and a decision-making timeline that can frustrate private sector partners and challenge internal change agents.

But this accountability serves a purpose. It protects against premature deployment, untested systems, vendor overpromises, and implementation failures. It forces rigor. It ensures that when you deploy

AI, you've thought through consequences, tested assumptions, and built safeguards.

Equity as a Core Requirement

Federal healthcare serves extraordinarily diverse populations. Active duty service members range from 18-year-old recruits to senior officers. Veterans span every demographic, geography, and socioeconomic category. Medicare and Medicaid beneficiaries include elderly populations, individuals with disabilities, low-income families, and children.

AI systems trained on narrow datasets, optimized for majority populations, or designed without equity considerations will fail in this environment. Worse, they'll fail unevenly—performing well for some populations while systematically disadvantaging others.

You cannot deploy AI that works well for some beneficiaries but poorly for others. You cannot accept algorithms that accurately diagnose conditions in some demographic groups while missing them in others. You cannot tolerate systems that streamline access for English speakers while creating barriers for non-English speakers.

Equity isn't a nice-to-have feature in federal AI. It's a functional requirement. Your AI systems must perform equitably across the populations you serve, or they're not mission-ready.

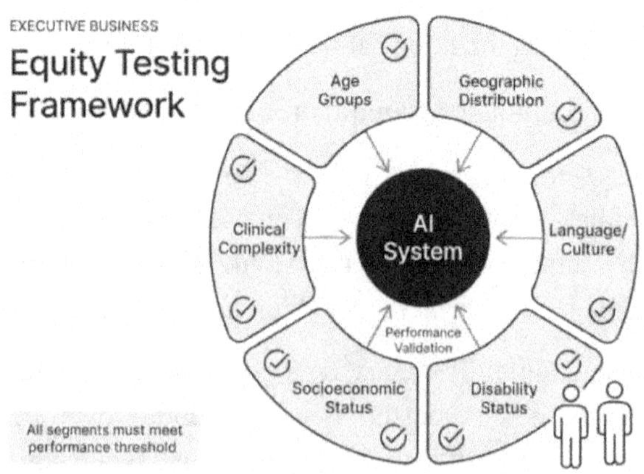

Constraints: Budget Cycles, Procurement Rules, and Oversight

Federal healthcare doesn't just operate differently—it operates under constraints that don't exist in commercial environments. These constraints aren't bureaucratic obstacles to overcome. They're the operating environment. Successful federal AI managers work within constraints, not against them.

2.5   Budget Cycle Realities

Private sector AI projects secure funding through venture capital, corporate budgets, or revenue allocation. Timelines are flexible. Priorities shift. Money flows to what's working.

Federal budgets operate on annual cycles, are planned years in advance, are appropriated by Congress, and are allocated through rigid processes. The money you'll use to deploy AI in fiscal year 2028 is being planned now. The decisions being made today about AI investments will constrain your options three years from now.

This creates predictability—you know funding timelines—but eliminates agility. You can't pivot quickly to a new AI opportunity. You can't scale a successful pilot without planning cycles in place. You can't shift resources from underperforming initiatives without formal reprogramming.

The implication: AI strategy in federal healthcare must align with budget cycles. Your AI roadmap isn't a dynamic document that evolves monthly—it's a multi-year plan that anticipates resource availability, procurement timelines, and appropriations realities.

Procurement Complexity

Acquiring AI capabilities in federal healthcare requires navigating the Federal Acquisition Regulation, agency-specific supplements, small-business set-asides, socioeconomic programs, and oversight requirements. You can't simply choose the best vendor and sign a contract.

Most AI vendors don't understand this environment. They're accustomed to commercial sales cycles: demo, proof of concept, negotiation, contract, deployment. Federal procurement involves:

- Market research and requirements definition

- Acquisition strategy development and approval

- Solicitation preparation and legal review

- Evaluation criteria and source selection planning

- Proposal evaluation and competitive range determination

- Negotiations and best and final offers

- Award protests and disputes

- Contract administration and oversight

This process takes months or years, not weeks. It requires specialized expertise—contracting officers, legal counsel, technical evaluators, and program managers. It demands documentation at every step.

The implication: You can't "move fast" in federal AI procurement. But you can move deliberately and effectively. Success requires understanding procurement pathways—other transaction authorities, blanket purchase agreements, indefinite delivery/indefinite quantity contracts, and strategic partnerships—and building acquisition strategies that comply with federal rules.

2.6   Oversight and Approval Hierarchies

Every significant AI initiative in federal healthcare will face multiple approval gates:

- Technical review boards evaluating architecture and security

- Clinical governance committees assessing safety and efficacy

- Privacy offices reviewing data handling and consent

- Security teams validating cybersecurity controls

- Legal offices ensuring regulatory compliance

- Budget offices confirming funding availability

- Leadership approval at multiple organizational levels

Each gate has a legitimate purpose. Technical reviews prevent deploying systems that won't integrate with the existing infrastructure. Clinical governance prevents unsafe implementations. Privacy reviews protect beneficiary rights. Security validation prevents breaches. Legal review ensures compliance. Budget oversight ensures fiscal responsibility.

But each gate adds time, requires documentation, and creates decision points where initiatives can stall.

The implication: Federal AI requires building approval to be built into your timeline. A 90-day pilot in concept becomes a 180-day effort when you account for approvals. A six-month deployment becomes a twelve-month program. This isn't inefficiency—it's the reality of operating in an environment where accountability requires oversight.

Federal AI Approval Pathway

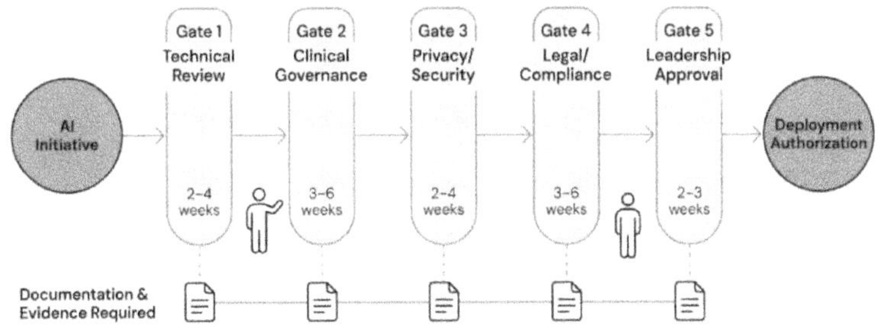

Plan AI timelines around governance, not just development.

The Innovation Paradox in Federal Agencies

Federal healthcare agencies are simultaneously among the most advanced and most constrained healthcare organizations worldwide.

DHA operates a global health system spanning continents. VA manages one of the largest integrated healthcare networks. CMS administers programs covering more than 150 million Americans.

These agencies employ world-class clinicians, operate sophisticated data infrastructure, and manage staggering complexity. Yet they often struggle to adopt innovations that smaller commercial organizations implement routinely.

This is the innovation paradox: organizational capacity exists, but organizational agility doesn't.

Why Innovation Stalls

Several factors create this paradox:

Legacy systems: Federal healthcare infrastructure includes systems built decades ago that run on outdated platforms and use obsolete protocols. These systems work—they support mission-critical operations—but they resist integration with modern AI tools. Replacing them is prohibitively expensive and operationally risky.

Risk aversion: Failed private sector initiatives become case studies. Failed federal initiatives become congressional hearings. This creates institutional risk aversion. Leaders who champion bold innovations face career risk if implementations fail. Safer to proceed cautiously, or not at all.

Workforce constraints: Federal healthcare organizations face hiring limitations, salary caps, and retention challenges. Recruiting AI talent is difficult when technology companies offer twice the salary. Building internal AI expertise is slow when training budgets are limited.

Change management complexity: Deploying AI in federal healthcare means coordinating across organizational silos, reconciling competing priorities, managing union relationships, addressing workforce concerns, and navigating political sensitivities.

A change that would take weeks in a private organization would take months or years in the federal context.

Vendor misalignment: Most AI vendors build for commercial markets—focusing on rapid deployment, cloud-native architecture, continuous updates, and usage-based pricing. Federal healthcare requires on-premise options, security certifications, stable releases, and fixed-price contracts. The mismatch creates friction.

Why Innovation Succeeds

Despite these challenges, federal healthcare agencies successfully deploy transformative AI when initiatives align with several success factors:

Clear mission connection: Innovations that demonstrably advance the mission get traction. Vague "transformation" initiatives stall. Specific solutions to concrete problems succeed.

Leadership sponsorship: AI initiatives need executive champions who provide top cover, remove barriers, and sustain momentum through approval processes. Without leadership sponsorship, even technically sound projects languish.

Incremental approach: Big-bang transformations fail. Incremental improvements succeed. Pilot projects that demonstrate value, then scale systematically, navigate organizational resistance more effectively than enterprise-wide rollouts.

Compliance by design: AI initiatives that integrate compliance requirements from inception succeed. Projects that treat compliance as an afterthought fail during approval processes or post-deployment audits.

Workforce engagement: Initiatives that engage frontline staff— clinicians, administrators, case workers—in design and testing succeed. Top-down technology mandates generate resistance and workarounds.

The implication: Federal AI success requires understanding both why innovation stalls and what drives its success. You're not fighting the system—you're learning to work effectively within it.

2.7   Why Federal Healthcare Is Uniquely Complex

Federal healthcare isn't a single system. It's an ecosystem of interconnected but distinct organizations, each with its own mission, population, authority, and constraints. Understanding this complexity is essential to deploying AI effectively.

Multiple Agencies, Different Missions

The Defense Health Agency serves active duty service members, maintaining medical readiness for global military operations. The priority is operational readiness—ensuring service members are medically fit for deployment.

The Department of Veterans Affairs serves former service members, providing lifetime healthcare for service-connected conditions and comprehensive care for eligible veterans. The priority is to honor service commitments and manage complex chronic conditions in aging populations.

The Centers for Medicare and Medicaid Services administer insurance programs for elderly Americans, individuals with disabilities, and low-income populations. The priority is to ensure access, protect beneficiaries, and steward public resources.

Department of Health and Human Services encompasses public health surveillance, epidemic response, regulatory oversight, and population health initiatives. The priority is to protect public health and safety across the nation.

Each agency operates under different statutory authorities, budget structures, oversight mechanisms, and stakeholder expectations. AI solutions that work brilliantly for one may be completely inappropriate for another.

The Clinical-Administrative Duality

Federal healthcare organizations are simultaneously clinical operations and administrative enterprises. They provide direct patient care and manage massive administrative processes. AI applications must address both dimensions.

Clinical AI encompasses decision support systems, diagnostic tools, risk-stratification models, and care-coordination platforms. These systems directly affect patient care, require clinical validation, demand high reliability, and fall under medical device regulations and clinical governance.

Administrative AI includes claims processing automation, scheduling optimization, supply chain management, fraud detection, and contact center augmentation. These systems affect operations, require business process validation, prioritize efficiency, and fall under operational governance and procurement rules.

The challenge: clinical and administrative AI have different stakeholders, approval processes, risk profiles, and success metrics. Managing both simultaneously requires navigating fundamentally different organizational cultures and decision-making structures.

Data Fragmentation and Interoperability Challenges

Federal healthcare generates massive volumes of data—electronic health records, claims data, pharmacy records, laboratory results, imaging studies, device data, and administrative transactions. This data sits in disparate systems using incompatible standards, managed by different organizations, and protected under varying security protocols.

DHA uses AHLTA and MHS GENESIS for electronic health records. VA uses VistA and is transitioning to Cerner. CMS receives claims data from thousands of providers using different billing systems. These systems don't natively communicate. Data exchange requires complex interfaces, manual reconciliation, and significant technical effort.

AI systems are data-hungry. They require large, clean, representative datasets for training and validation. When data is fragmented across incompatible systems, building those datasets becomes a major undertaking—often consuming more time and resources than the AI development itself.

The implication: Federal healthcare AI projects must account for the complexity of data integration. Your AI roadmap must include data strategy—how you'll access, aggregate, standardize, and govern the data that AI systems require.

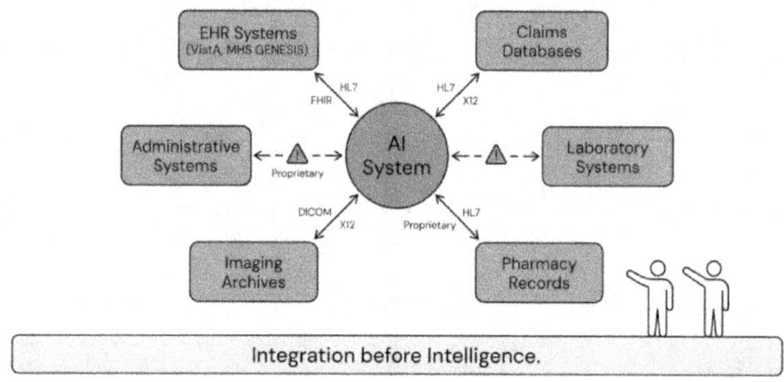

Contractor and Vendor Ecosystem

Federal healthcare doesn't build all its technology in-house. Much capability comes through contractors—system integrators, technology vendors, managed service providers, consulting firms. This creates a complex ecosystem in which AI deployment requires coordinating among government program managers, prime contractors, subcontractors, and technology vendors.

Contract structures shape what's possible. Performance-based contracts incentivize outcomes. Time-and-materials contracts pay for effort. Fixed-price contracts lock in scope. Indefinite delivery/indefinite quantity contracts provide flexibility. Each structure enables certain AI approaches and constrains others.

Vendor relationships matter. Established contractors understand federal requirements, have security clearances, know approval processes, and have past performance records. New vendors bring innovation but lack federal experience. Managing this ecosystem— balancing innovation and reliability, new entrants and incumbents, competition and continuity—is core to federal AI strategy.

2.8   The Compliance-by-Design Mindset

The single most important shift for federal healthcare AI managers: moving from compliance as a checkpoint to compliance as a design principle.

In commercial AI, compliance often comes late. Build the system, demonstrate value, then address regulatory requirements. This sequence works when regulations are limited, and enforcement is probabilistic.

In federal healthcare, this approach fails. AI systems must comply with:

- Executive orders on AI governance and safety

- Office of Management and Budget requirements

- National Institute of Standards and Technology AI Risk Management Framework

- Health Insurance Portability and Accountability Act privacy and security rules

- Federal Information Security Management Act requirements

- Federal Risk and Authorization Management Program for cloud services

- Agency-specific policies and standards

Compliance isn't a final gate—it's a continuous requirement woven through design, development, testing, deployment, and operations. Systems that don't embed compliance from inception won't pass approval processes. Systems that achieve compliance at launch but don't maintain it will face audit findings, operational restrictions, or shutdown.

What Compliance-by-Design Means Practically

Compliance-by-design transforms how you approach AI:

Requirements phase: Compliance requirements aren't separate from functional requirements—they're integrated. When defining what an AI system must do, you simultaneously define what it must not do, what data it can access, what decisions require human approval, what audit trails it must create, and what performance thresholds it must maintain.

Vendor evaluation: Vendor claims aren't evaluated purely on capability—they're evaluated on compliance readiness. Can the vendor demonstrate HIPAA compliance? Do they have FedRAMP authorization? Can they provide security documentation? Will they accept federal contract terms? Can they support continuous monitoring?

Testing and validation: Validation isn't just about accuracy—it's about fairness, explainability, security, and reliability. You test AI systems for bias across demographic groups. You validate that explanations are meaningful. You probe for adversarial vulnerabilities. You verify that performance remains stable over time.

Deployment: Launch isn't about speed—it's about readiness. You deploy when security controls are operational, monitoring infrastructure is functioning, staff are trained, escalation procedures are documented, and approval authorities have signed off.

Operations: Production isn't static—it's continuously validated. You monitor AI performance, detect drift, review decisions, audit outcomes, update models, and document changes.

This compliance-by-design mindset doesn't slow innovation—it channels innovation toward solutions that can actually be deployed, sustained, and trusted in federal healthcare environments.

2.9   What This Means for You on Monday Morning

Understanding why federal AI is different isn't academic—it shapes every decision you make as a manager. When you return to your desk, these principles translate into concrete actions:

Evaluate AI opportunities through a mission lens. When vendors pitch solutions, your first filter is mission impact. When staff proposes pilots, your first question is mission value. When leadership asks about the AI strategy, your answer centers on advancing the mission.

Build compliance into every conversation. Compliance isn't a separate workstream—it's integrated into requirements, vendor evaluation, testing, and operations. You don't ask "Can we do this?" followed by "Can we do it compliantly?" You ask, "How do we do this compliantly?" from the start.

Plan for constraint, not around it. Budget cycles, procurement timelines, and approval processes aren't obstacles to overcome— they're the environment you operate in. Your AI roadmap accounts for these realities. Your schedules reflect approval timelines. Your strategies work within acquisition rules.

Invest in transparency and explainability. Black-box AI doesn't survive federal scrutiny. You prioritize systems that can explain their reasoning, provide audit trails, and support appeals. You build documentation from day one.

Design for equity from inception. You don't deploy AI and then check for bias—you test for equitable performance across populations throughout development. Equity is a functional requirement, not a compliance checkbox.

Communicate in mission terms. You translate technical concepts into mission language. You explain AI initiatives in terms of readiness, beneficiary care, program integrity, and public health outcomes. You help leadership and stakeholders understand the value of AI in terms they care about.

Build incrementally. You don't propose enterprise-wide AI transformation. You identify high-value use cases, pilot systematically, validate thoroughly, and scale deliberately. You demonstrate value before seeking scale.

These aren't aspirational principles—they're operational necessities. Federal healthcare AI succeeds when managers operate within these realities rather than against them.

2.10 Moving Forward

Federal healthcare AI is different—profoundly different. Different mission, different constraints, different accountability structures, different risk tolerances, different success criteria. Approaches that work in commercial environments fail here. Strategies that succeed here look nothing like Silicon Valley playbooks.

This isn't a limitation—it's a clarification. Understanding what makes federal AI different allows you to deploy it effectively. You're not trying to import commercial approaches wholesale. You're adapting proven AI capabilities to serve federal missions, operate within federal constraints, meet federal accountability standards, and honor federal commitments to service members, veterans, beneficiaries, and citizens.

The chapters that follow build on this foundation. They explore the specific agencies, use cases, implementation pathways, procurement strategies, governance models, and practical tools you need to deploy AI successfully in federal healthcare. Each chapter assumes you understand why federal AI is different—and provides specific guidance on what works in this unique environment.

You're not starting from zero. Federal healthcare organizations have successfully deployed AI—improving care delivery, accelerating claims processing, detecting fraud, optimizing operations, and advancing their missions. These successes share common patterns: clear mission alignment, leadership sponsorship, compliance by design, incremental deployment, and workforce engagement.

Your opportunity is to learn from these successes, apply these patterns, avoid known pitfalls, and build AI capabilities that serve the mission while maintaining the public trust that makes federal healthcare possible.

The work begins with understanding that federal AI is different. It continues with learning how to make it work.

2.11 Chapter Summary

Federal healthcare AI operates in a fundamentally different environment than commercial AI. Five key differences shape everything:

Mission primacy: AI serves mission, not markets. Value is measured in readiness, beneficiary care, program integrity, and public health outcomes—not revenue or market share. Risk is calculated against mission consequences. The social contract requires transparency.

Accountability environment: Federal managers operate under oversight by agency leadership, Congress, inspectors general, GAO, the media, advocacy groups, and the public. This creates documentation requirements, approval hierarchies, and timelines that demand rigor.

Operational constraints: Budget cycles, procurement regulations, and approval processes define the operating environment. These aren't obstacles—they're the reality you work within. Success requires aligning AI strategy with these constraints.

Unique complexity: Federal healthcare is an ecosystem of agencies with different missions, populations, and authorities. Data is fragmented. Systems don't natively interoperate. Clinical and administrative operations require different approaches. Contractor relationships shape implementation options.

Compliance as design principle: Compliance-by-design means integrating regulatory requirements, security controls, privacy protections, and oversight mechanisms from inception—not bolting them on afterward. This mindset transforms how you approach requirements, vendor evaluation, testing, deployment, and operations.

Federal healthcare AI succeeds when managers understand these differences and operate accordingly—evaluating opportunities through a mission lens, building compliance into every decision, planning within constraints, prioritizing transparency, designing for equity, communicating in mission terms, and building incrementally.

The approach that works isn't about moving fast—it's about moving deliberately and effectively. It's about deploying AI that serves federal missions, operates within federal constraints, meets federal accountability standards, and honors federal commitments.

That's the foundation. The rest of this book builds the operational playbook.

3 The Federal Healthcare Ecosystem

A program manager at the Department of Veterans Affairs just received approval to pilot an AI tool that summarizes complex patient charts into concise clinical briefs. The pilot succeeds at three VA medical centers in the Mid-Atlantic region. Leadership asks a reasonable question: Can we scale this across all 1,380 VA facilities? The answer is not what anyone expects. Some facilities still run VistA. Others have migrated to the new federal electronic health record. Data formats differ. Governance approvals differ. Workforce readiness differs. The pilot that worked beautifully in three locations now faces an ecosystem of constraints that no single team anticipated.

This scenario plays out constantly across federal healthcare. It explains why the Department of Veterans Affairs retired 72 AI use cases in a single inventory cycle—not because the technology failed, but because the ecosystem wasn't ready. It explains why the Defense Health Agency can deploy AI globally for combat casualty care but struggles to integrate the same tools with stateside administrative systems. It explains why CMS can analyze a billion claims but can't easily share insights with VA or DHA.

Federal healthcare isn't a single system. It's an ecosystem—a complex network of agencies, each with distinct missions, populations, governance structures, data architectures, and operational models. Understanding this ecosystem isn't background reading. It's an operational necessity.

3.1 Why This Matters

Every AI deployment decision you make is shaped by the ecosystem you operate within. Choosing the wrong procurement vehicle wastes months. Building on the wrong data platform wastes millions. Ignoring interagency data-sharing constraints kills promising pilots. Understanding the ecosystem—who operates in it, what drives their decisions, where integration opportunities exist, and what constraints you'll face—is the difference between AI that delivers mission value and AI that becomes a cautionary tale in the next Inspector General report.

This chapter maps the federal healthcare ecosystem in operational detail. It examines the major organizational players—Defense Health Agency, Veterans Health Administration, Centers for Medicare and Medicaid Services, the broader Department of Health and Human Services, and the Indian Health Service. It explores how these agencies interoperate, where they don't, why data fragmentation persists, how contractors shape implementation, and why the clinical-administrative duality matters for every AI initiative you lead.

3.2   Defense Health Agency: The Military Health System

The Defense Health Agency operates the Military Health System, serving 9.6 million beneficiaries worldwide. But describing DHA as a healthcare organization misses its fundamental nature: DHA is a combat support agency. Everything it does—clinical care, medical logistics, research, information systems—exists to generate and sustain warfighter medical readiness.

This mission distinction shapes every AI conversation. When evaluating AI opportunities in DHA, the first question isn't "Does this improve healthcare delivery?" It's "Does this enhance warfighter readiness and medical capability in support of military operations?"

Mission Architecture and Organizational Structure

DHA operates through nine Defense Health Networks—regional structures that deliver healthcare across military hospitals and clinics worldwide. These networks aren't just administrative divisions. They're operational structures designed to ensure medical readiness across geographic areas spanning continents, time zones, and operational environments.

The DHA organizational model reflects military command structures more than civilian healthcare administration. The agency director reports to the Assistant Secretary of Defense for Health Affairs. Beneath that, DHA operates through functional offices—Healthcare Operations, Medical Affairs, Health Care Administration, Information Operations—that provide enterprise-level capabilities to the nine regional networks.

This structure creates both opportunities and constraints for AI deployment. The enterprise functional offices can drive standardization—one AI capability deployed across all networks, ensuring consistency and interoperability. But this same centralization creates approval hierarchies, requires multi-level coordination, and extends implementation timelines.

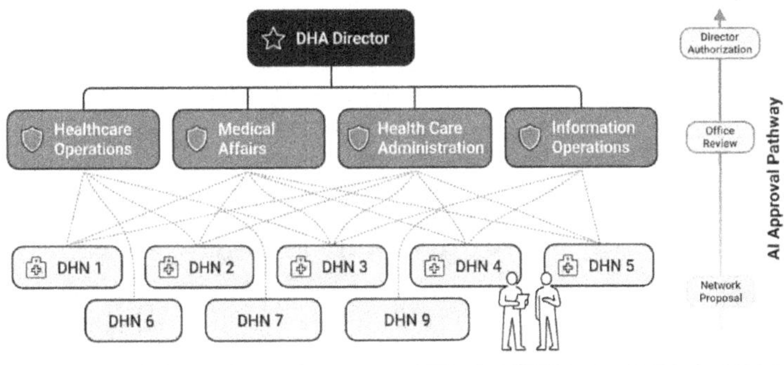

AI ideas move from Defense Health Networks through DHA offices to Director-level authorization.

The Readiness Imperative

Readiness isn't an abstract concept in DHA—it's a measurable operational requirement. Medical readiness means service members are medically qualified to deploy. Medical warrior currency means military healthcare providers maintain the clinical skills necessary to deliver care in operational environments. These requirements drive AI priorities in ways commercial healthcare has never faced.

AI applications in DHA must support one or more readiness missions:

Warfighter Medical Readiness: Ensuring service members are medically fit for deployment through predictive analytics, automated health assessments, and accelerated medical clearance processes.

Medical Warrior Currency: Maintaining provider clinical skills through case volume optimization, surgical scheduling AI, and skills maintenance tracking systems.

Joint Warfighting Capabilities: Supporting operational medicine through medical logistics AI, trauma system optimization, blood supply chain management, and deployment health surveillance.

Healthcare Delivery: Providing timely, high-quality care to beneficiaries through clinical decision support, appointment optimization, and care coordination platforms.

When vendors pitch AI solutions to DHA, they often emphasize cost savings or efficiency gains. These matter, but they're secondary. The primary evaluation criterion is the impact on readiness. An AI system that reduces costs but doesn't enhance readiness has limited value. An AI system that costs more but significantly improves readiness may be mission-critical.

3.3   The Global Deployment Reality

DHA operates globally—fixed facilities in the continental United States, overseas military hospitals in Europe and Asia, forward-deployed medical units in operational theaters, and expeditionary capabilities that move with military operations. This creates AI deployment challenges that don't exist in domestic commercial healthcare.

Network connectivity varies dramatically. A military hospital in Germany has reliable high-bandwidth connections. A forward surgical team in an operational environment may have limited, intermittent, or tactically restricted connectivity. AI systems designed for always-on cloud environments don't work in bandwidth-constrained or communications-denied settings. This drives specific technical requirements:

Edge computing capabilities: AI systems must operate locally when cloud connectivity is unavailable.

Disconnected operations: Systems must handle intermittent connectivity, synchronize when connections are restored, and maintain functionality during communications blackouts.

Bandwidth efficiency: Models must be optimized for low-bandwidth environments, using compressed data transmission and efficient update mechanisms.

Security in austere environments: AI systems must maintain security controls even in forward-deployed settings where physical and network security may be limited.

MHS GENESIS and the Data Modernization Inflection Point

DHA's transition to MHS GENESIS—the integrated electronic health record built on Oracle Cerner architecture and shared with VA—represents the most significant health IT modernization in federal healthcare history. MHS GENESIS provides a common data platform across DHA facilities, creating unprecedented AI opportunities: standardized data structures, consistent clinical workflows, uniform documentation practices, and seamless data access.

In practice, the transition creates temporary data fragmentation as facilities migrate at different times. The AI implications are significant:

Data availability timing: AI initiatives must account for which facilities have migrated and which remain on legacy systems. Training models on data from migrated facilities, while others use legacy systems, creates version control challenges.

Workflow standardization: MHS GENESIS enforces standardized clinical workflows with clear AI integration points. But customizations built into legacy systems may not translate.

Interoperability with VA: The shared platform creates opportunities for joint data. AI models trained on combined DHA-VA data leverage larger datasets. But data sharing requires governance agreements, privacy protocols, and security controls.

Vendor ecosystem constraints: Third-party AI vendors must integrate with the Oracle Cerner architecture. Integration capabilities and API availability shape what's technically feasible.

MHS GENESIS Migration and AI Data Strategy

Align AI roadmaps with the MHS GENESIS migration timeline.

Veterans Health Administration: Lifetime Care at Scale

The Veterans Health Administration operates the largest integrated healthcare system in the United States—1,380 healthcare facilities serving 9.1 million enrolled veterans annually. With 371,000 employees and an annual budget exceeding $ 68 billion, VHA delivers comprehensive healthcare across the full spectrum of needs—from routine primary care to specialized services for traumatic injuries, mental health conditions, and complex chronic diseases.

VA's mission differs fundamentally from DHA's focus on operational readiness. VA honors the nation's commitment to veterans by providing lifetime healthcare for service-connected conditions and comprehensive care for eligible veterans. VA also has the most mature AI portfolio in federal healthcare—367 cataloged use cases as of 2025, more than any other federal health agency.

3.4   Organizational Structure: The VISN Model

VHA operates through 22 Veterans Integrated Service Networks—regional structures that manage medical centers, outpatient clinics, and community-based facilities. Each VISN functions with significant operational autonomy, managing budgets based on enrolled veteran populations and adapting services to regional demographics.

This decentralized model creates both AI opportunities and challenges:

Pilot flexibility: VISNs can initiate AI pilots tailored to local needs without national-level approval, enabling faster innovation cycles and localized testing.

Scaling complexity: Successfully scaling a VISN pilot to national deployment requires different approval processes, enterprise governance, and coordination across 22 semi-autonomous networks.

Inconsistent adoption: AI capabilities may achieve adoption in some VISNs while others resist or prioritize alternatives, creating fragmented veteran experiences.

Knowledge-sharing gaps: Best practices in one VISN may not systematically transfer to others without formal knowledge-management mechanisms.

3.5   The Veteran Population: Complexity That Shapes AI

VA serves an extraordinarily diverse population. Veterans range from young adults recently separated from service to elderly World War II and Korean War veterans. They span every demographic, socioeconomic status, geographic location, and clinical complexity level. Many carry service-connected conditions requiring specialized long-term care. Others face mental health challenges, including post-traumatic stress disorder and traumatic brain injury.

This diversity creates specific AI requirements:

Performance across demographics: AI must perform equitably for young and older veterans, male and female veterans, veterans of diverse racial and ethnic backgrounds, and across different service eras.

Clinical complexity handling: Models must account for polytrauma, multiple comorbidities, service-connected conditions, and interactions between military injuries and age-related diseases.

Geographic variation: AI must work in urban medical centers with full specialty services and rural clinics with limited resources.

Care continuity across transitions: Veterans move between military healthcare, VA, community care, and Medicare. AI must handle fragmented histories and incomplete records.

3.6   VA's AI Portfolio: What 367 Use Cases Reveal

VA's 2025 AI Use Case Inventory provides the most detailed picture of federal healthcare AI deployment anywhere in government. The inventory catalogs 367 AI use cases: 253 in healthcare delivery, 27 in benefits processing, 57 in IT operations, and 30 across other mission areas. Of these, 138 are actively deployed in production, 136 are in pre-deployment, and 72 have been retired.

AI That Works: Proven VHA Deployments

STORM—Stratification Tool for Opioid Risk Mitigation. This AI-driven clinical decision support tool flags veterans at elevated risk of overdose or suicide related to opioid use. It uses electronic health records and pharmacy data to generate a risk score for each patient on long-term opioid therapy, surfacing high-risk cases to care teams. STORM doesn't replace clinician judgment—it prioritizes which patients should receive case review, care plan adjustments, naloxone co-prescribing, or specialty referral. Use of STORM has been associated with a roughly 22 percent decrease in mortality among high-risk opioid patients when sites followed required case review steps. It shifts opioid safety from ad-hoc chart review to a systematic, data-driven process.

AI-Assisted Colonoscopy. VHA sites have deployed FDA-cleared AI devices that use real-time computer vision to highlight potential polyps during colonoscopy. The AI acts as a "second observer," highlighting suspicious lesions so endoscopists can inspect them more closely. The device integrates into existing endoscopy workflows with no changes to scheduling, prep, or documentation. A VA study showed a statistically significant 21 percent increase in the odds of adenoma detection—a direct cancer-prevention outcome.

AI-Assisted Clinical Documentation. VHA authorized internal generative AI tools—VA GPT and Microsoft 365 Copilot Chat—to help clinicians draft documentation, summarize complex charts, and generate patient instructions. Over 95,000 VA employees adopted VA GPT, with users reporting two to three hours saved per week and more than 70 percent reporting improved job satisfaction. Output is treated as a draft—clinicians verify accuracy before anything is entered into the record.

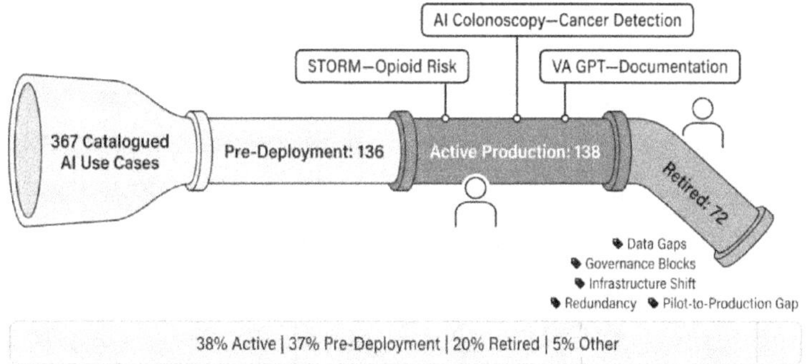

VA AI Deployment Maturity Spectrum

AI That Failed: What 72 Retirements Reveal

The 72 retired use cases aren't public failures in the traditional sense. VA doesn't publish individual retirement rationales. But the pattern reveals five systemic deployment pitfalls instructive for every federal healthcare AI initiative.

Data Readiness Gaps. AI models trained on one system's data don't translate cleanly to another during multi-year EHR transitions. The Government Accountability Office found that VA failed to ensure that data transferred during the EHR rollout met clinicians' needs—clinicians couldn't access patients' allergies, medications, and immunizations in the new system. Pilots built on incomplete or siloed data can't survive production environments where data quality differs dramatically from the controlled pilot setting.

Governance and Compliance Blocks. Federal surveys indicate that 45 percent of stalled AI projects cite governance and compliance as the primary barrier. Only half of agencies have a formal AI governance framework, and just 41 percent have a dedicated AI budget. At VA, the Inspector General found VHA does not have a formal mechanism to identify, track, or resolve risks associated with generative AI, and that key patient safety offices were not consulted before AI tools were authorized for clinical use.

Pilot-to-Production "Valley of Death." Between fiscal years 2020 and 2024, the federal government invested approximately $ 15 billion in AI—but 86 percent went to research and development, while only 14 percent funded operational deployment. Eighty-one percent of federal leaders admit they are more likely to launch a new pilot than expand an existing one. Pilots operate with controlled data, small user bases, and relaxed governance. Production requires enterprise integration, security certification, change management, and sustained funding.

Consolidation of Redundant Efforts. In a department with 367 cataloged use cases, overlap is inevitable. Multiple teams independently build NLP summarization tools, document classifiers, or risk-prediction models. Retiring duplicates frees resources for the winners. This is healthy portfolio management, not failure—but it reflects insufficient coordination before pilots launch.

Misalignment with Evolving Infrastructure. The ongoing EHR modernization fundamentally changes the technical baseline. AI tools designed for VistA workflows become orphaned as facilities migrate to the new system. VA's updated AI strategy acknowledged this: as AI tools are validated and show worth, they will be incorporated into the EHR and other platforms. Tools that can't bridge the transition get retired regardless of technical merit.

3.7   The VistA Legacy and Modernization Journey

VA's Veterans Health Information Systems and Technology Architecture (VistA) has served as the backbone of VA healthcare IT for decades. VistA pioneered integrated electronic health records, clinical decision support, and computerized provider order entry. But VistA's architecture dates to the 1980s, relies on outdated technology stacks, and resists integration with modern AI tools.

VA is transitioning from VistA to the same Oracle Cerner platform used by DHA—creating the joint Federal Electronic Health Record. The transition was paused in April 2023 due to patient safety concerns and only resumed in 2025 at select facilities. During transition, AI initiatives must navigate:

- Dual system operations: Some facilities run VistA, others the new EHR, some both. AI must integrate with both or accept limited deployment.

- Data migration timing: Historical VistA data migrates over time. AI requiring longitudinal data must access both systems or accept incomplete histories.

- Workflow disruption: The transition changes clinical workflows, documentation practices, and data structures. AI designed for VistA may not match new EHR workflows.

- Vendor relationship shifts: Vendors with VistA integration expertise must develop new capabilities or risk obsolescence.

Benefits Administration and Claims Processing

Beyond healthcare delivery, VA manages massive benefits administration—disability compensation claims, pension determinations, appeals processing, and benefits verification. Claims processing AI must handle unstructured data—medical evidence, service records, disability evaluations, written statements—while applying complex regulatory criteria. Incorrect determinations delay veterans' access to earned benefits; insufficient review may result in benefits being granted inappropriately. The balance between efficiency and accuracy shapes VA's entire administrative AI strategy.

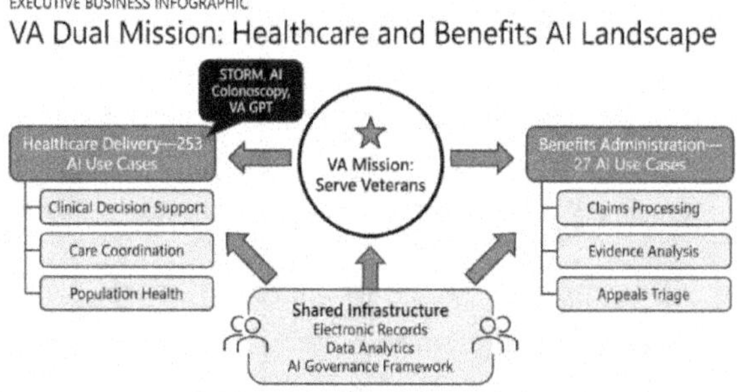

Centers for Medicare and Medicaid Services: Population-Scale Healthcare

CMS operates at a fundamentally different scale and model than DHA or VA. Rather than directly delivering healthcare through government-owned facilities, CMS administers insurance programs covering over 150 million Americans—elderly Medicare beneficiaries, individuals with disabilities, low-income Medicaid recipients, and children in the Children's Health Insurance Program.

This indirect delivery model creates unique opportunities and constraints for AI. CMS doesn't control clinical workflows in tens of thousands of independent practices, hospitals, and health systems. Instead, CMS influences behavior through payment policy, quality measurement, fraud detection, and regulatory oversight. AI serves these levers at a scale no other organization approaches.

Organizational Structure and Program Administration

CMS operates through several major centers, each with distinct AI priorities:

- Center for Medicare: Administers Parts A, B, C, and D, covering hospital insurance, medical insurance, Medicare Advantage, and prescription drug coverage for 67 million beneficiaries.

- Center for Medicaid and CHIP Services: Oversees state-administered Medicaid programs and CHIP, providing federal guidance, funding, and oversight for 90 million enrollees.

- Center for Program Integrity: Detects and prevents fraud, waste, and abuse through data analytics, provider screening, payment review, and law enforcement coordination.

- Center for Clinical Standards and Quality: Develops quality measures, administers quality reporting programs, certifies providers, and operates survey processes.

- Center for Innovation: Tests new payment and service delivery models, evaluates innovation initiatives, and scales successful approaches.

For AI managers, this structure means different CMS centers have different AI priorities, data access authorities, approval processes, and stakeholder relationships. An AI initiative for fraud detection operates under different governance than one for quality measurement.

3.8 The Indirect Delivery Model and Data Ecosystem

CMS receives data rather than generating it directly. Claims data flows from tens of thousands of providers. Enrollment data comes from the Social Security Administration and state Medicaid agencies.

Quality data arrives through standardized reporting. This creates an extraordinarily rich but complex data environment.

- Massive scale: Billions of claims annually, petabytes of historical data, millions of providers, and hundreds of millions of beneficiary records create computational challenges requiring specialized infrastructure.

- Standardized but inconsistent: Claims use standard formats—ICD, CPT, HCPCS—but coding practices vary across providers, regions, and specialties. AI must handle variation without mistaking coding differences for clinical differences.

- Structured transaction data: CMS primarily receives structured claims transactions, simplifying some AI applications but limiting clinical detail available for model training.

- Time lags: Claims arrive weeks or months after services. Real-time AI faces inherent latency constraints.

- Limited clinical detail: Claims capture what was billed, not detailed clinical information. AI must infer clinical patterns from administrative data or integrate supplementary sources.

3.9 Program Integrity and Fraud Detection

Protecting Medicare and Medicaid from fraud, waste, and abuse is among the highest-value AI applications in federal healthcare. With annual expenditures exceeding one trillion dollars, even small percentages of improper payments translate to billions. AI enhances program integrity through pattern recognition, anomaly detection, network analysis, and predictive modeling.

- False positive consequences: Incorrectly flagging legitimate providers damages relationships, delays payments, and creates an administrative burden. Precision matters as much as recall.

- Adversarial adaptation: Fraudulent actors adapt to detection methods. Models must continuously evolve to detect new schemes while maintaining the ability to detect known patterns.

- Investigation capacity: AI can identify thousands of suspicious cases, but its investigation capacity is limited. Models must prioritize by likelihood and recovery potential.

- Legal evidentiary standards: Fraud cases require evidence meeting legal standards. AI detections must support—not replace—investigative due diligence.

- Due process requirements: Providers have rights to understand why they're flagged, appeal determinations, and receive due process. Black-box AI is incompatible.

State-Federal Partnership in Medicaid

Medicaid's state-federal partnership creates unique complexities in AI. CMS provides federal funding and oversight, but states design benefits, determine eligibility, set payment rates, and administer operations. Fifty states plus territories operate programs with significant variation in technical sophistication and AI readiness.

- State autonomy: States control Medicaid AI decisions. CMS can encourage or incentivize but cannot unilaterally deploy AI across state programs.

- Capability variation: State agencies vary dramatically in technical sophistication. Approaches feasible in large states may be impractical in smaller ones.

- Federal approval requirements: State AI affecting eligibility, benefits, or funding typically requires CMS approval through waiver or state plan amendment processes.

- Cross-state learning: Successful pilots in one state should inform others, but knowledge transfer mechanisms remain informal.

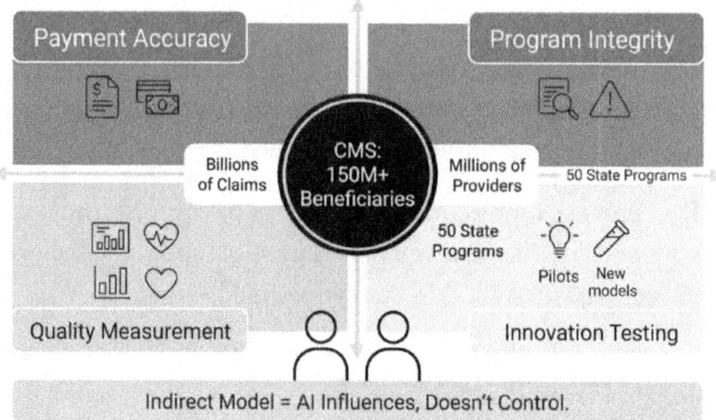

Department of Health and Human Services: The Broader Ecosystem

HHS encompasses far more than CMS. Multiple agencies play roles in federal healthcare AI—setting policy, conducting research, regulating products, monitoring public health, and overseeing compliance.

3.10 Key HHS Agencies and Their AI Roles

- Food and Drug Administration: Regulates AI-enabled medical devices, software as a medical device, and clinical decision support. FDA approval or clearance may be required for certain clinical AI—a critical gate shaping vendor selection and timelines.

- Centers for Disease Control and Prevention: Deploys AI for outbreak detection, epidemiological modeling, and health threat assessment. CDC data and capabilities inform broader federal health AI.

- National Institutes of Health: Funds biomedical AI research generating models, datasets, and methodologies that federal agencies can leverage.

- Office for Civil Rights: Enforces HIPAA privacy and security rules. OCR guidance shapes how AI systems handle protected health information.

- Office of the National Coordinator for Health IT: Promotes interoperability, certifies health IT products, and develops data exchange standards.

- Indian Health Service: Provides healthcare to American Indians and Alaska Natives through 605 facilities, facing unique AI challenges in geographically dispersed, resource-constrained settings.

For AI managers, these HHS organizations aren't peripheral. FDA determines whether your clinical AI requires regulatory approval. OCR determines whether your data handling is lawful. ONC determines which interoperability standards your systems must support. Ignoring these stakeholders creates compliance risks that surface at the worst possible time.

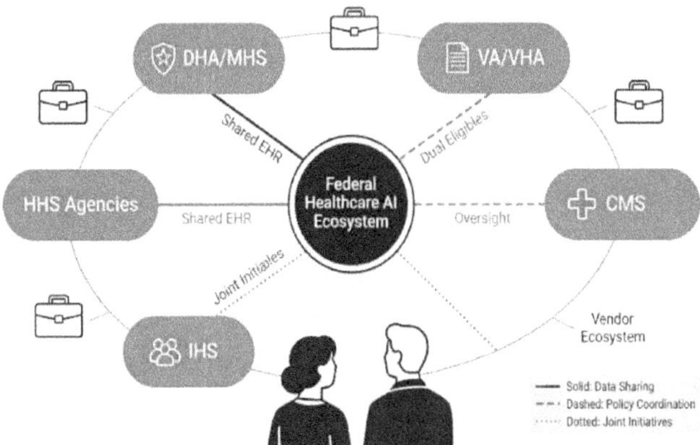

How These Agencies Interoperate—and Where They Don't

Federal healthcare agencies operate as semi-independent entities, each with statutory authorities, dedicated budgets, separate IT systems, and distinct governance structures. This independence serves important purposes but creates interoperability challenges that directly affect AI deployment.

3.11   Care Continuity Across Agency Boundaries

Beneficiaries often move between federal healthcare systems. Service members transition from DHA to VA after military service. Veterans become Medicare-eligible at 65 and can receive care through both VA and Medicare providers. Approximately 12 million dual-eligible individuals receive benefits from both Medicare and Medicaid. AI systems that can't see across boundaries can't serve patients who cross them.

- Incomplete health histories: When individuals transition, complete records don't automatically transfer. AI requiring longitudinal data faces gaps that degrade predictions.

- Duplicate testing: Without cross-system visibility, providers repeat tests. AI designed to reduce unnecessary utilization can't optimize across boundaries.

- Medication reconciliation: Individuals with prescriptions from multiple systems face safety risks. AI medication safety tools need complete medication lists.

- Care coordination complexity: Coordinating care for complex patients across systems requires information sharing that the current infrastructure doesn't fully support.

Data Sharing Agreements and Governance

Sharing health data across agencies requires formal agreements addressing legal authorities, privacy protections, security requirements, permitted uses, and governance. These take time but enable multi-agency AI applications.

- Memoranda of Understanding: Formal agreements defining data sharing purposes, uses, security, and governance. AI needing cross-agency data typically requires MOUs—a process measured in months.

- Health information exchanges: Technical infrastructure for electronic health data exchange. HIEs support care continuity but may limit access to bulk data for AI model development.

- Research data sharing: Agencies share de-identified data for research under established frameworks, subject to approval and data use agreements.

- Statutory authorities: Some laws mandate or authorize specific sharing. Understanding these clarifies what's permissible versus what needs new agreements.

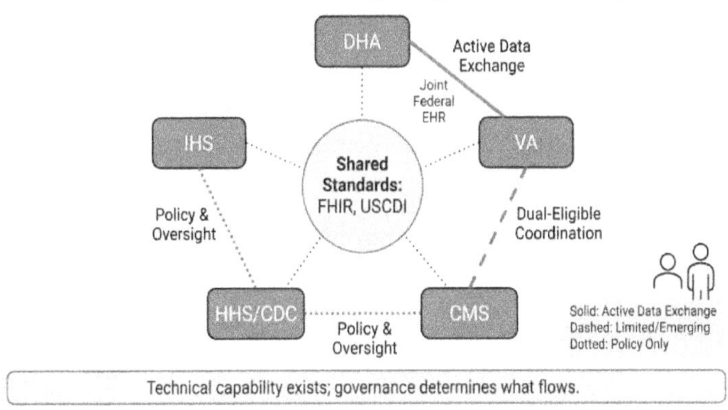

Data Fragmentation and Interoperability Challenges

Federal healthcare generates massive data volumes across disparate systems using incompatible standards, managed by different organizations, and protected under varying protocols. This fragmentation represents the single largest barrier to federal healthcare AI at scale, operating on three compounding dimensions.

Technical Fragmentation

- Incompatible EHR platforms: DHA uses MHS GENESIS on Oracle Cerner. VA transitions from VistA to Cerner. Thousands of Medicare/Medicaid providers use Epic, Allscripts, Meditech, and hundreds of others.

- Data standard variation: HL7 v2 messaging, HL7 FHIR APIs, C-CDA documents, DICOM imaging, X12 claims transactions—different systems support different standards at different maturity levels.

- Terminology inconsistency: SNOMED CT for clinical terms, ICD for diagnoses, CPT for procedures, LOINC for labs, RxNorm for medications. Mapping between terminologies complicates AI NLP.

- Legacy system persistence: Decades-old systems with outdated protocols, limited APIs, and batch-oriented access. Integrating real-time AI creates fundamental architectural challenges.

Semantic Fragmentation

- Documentation variation: Primary care physicians, specialists, and nurses document differently. Some provide narratives; others use brief structured notes. AI NLP must handle this range.

- Coding inconsistency: Same conditions coded differently depending on the coder's training and institutional practices.

- Abbreviation ambiguity: Thousands of abbreviations, many context-dependent. "MS" could mean multiple sclerosis, mitral stenosis, or mental status.

- Negation and uncertainty: "No evidence of infection" and "possible pneumonia" require sophisticated language understanding for correct extraction.

Organizational Fragmentation

- Data ownership boundaries: Each agency controls access. Cross-agency AI requires negotiating agreements addressing legal authorities, privacy, uses, and security.

- Privacy interpretation variation: Agencies interpret HIPAA differently and implement different controls. Harmonizing for joint AI requires extensive coordination.

- Competing priorities: Sharing that advances one agency's mission may be tangential to another's. Alignment requires demonstrating mutual benefit.

- Trust dynamics: Historical relationships and institutional cultures affect willingness to collaborate on AI.

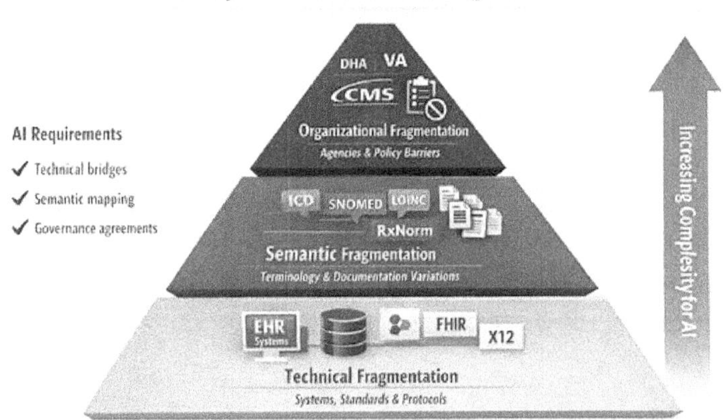

Interoperability Progress and Emerging Standards

Despite fragmentation, significant progress creates real AI opportunities:

- FHIR adoption: Modern APIs for health data exchange. Federal agencies increasingly support FHIR, enabling standardized access to AI data—though adoption remains incomplete across legacy systems.

- USCDI: Standardized health data elements that systems must support. AI designed around USCDI has clearer expectations for data availability.

- 21st Century Cures Act: Information blocking provisions improve data access for AI initiatives.

- Joint federal EHR: The shared DHA-VA platform creates unprecedented interoperability between military and veteran healthcare.

3.12 The Role of Contractors, Integrators, and Vendors

Federal healthcare doesn't build all capabilities in-house. A complex ecosystem of contractors, system integrators, technology vendors, and managed service providers delivers significant portions of AI capabilities. Understanding this ecosystem is essential—misunderstanding it is one of the fastest paths to failed deployments.

The Contractor Landscape

- Prime contractors and system integrators: Large firms winning major contracts and managing overall delivery. They integrate multiple vendors, manage subcontractors, and serve as the primary government interface.

- Technology vendors: Companies providing specific AI capabilities or platforms. Some specialize in federal markets; others enter from commercial space with strong technology but a weak understanding of federal constraints.

- Managed service providers: Firms operating IT infrastructure and applications under long-term service contracts.

- Small businesses and startups: Innovative firms bringing cutting-edge AI. They benefit from set-aside programs but may lack federal contracting experience, often partnering with larger primes.

- R&D contractors: Organizations conducting applied research and developing prototypes before technologies transition to operational use.

How Contract Structures Shape AI

- Performance-based contracts: Pay for outcomes. These incentivize AI performance and mission impact, but defining measurable AI outcomes requires sophisticated metrics frameworks.

- Time-and-materials: Pay for hours. Flexible for exploratory AI work, but doesn't incentivize efficiency or outcomes.

- Firm-fixed-price: Locked scope and price. Works when requirements are well-defined; risky for innovative applications where feasibility may shift.

- IDIQs: Framework for ordering multiple projects from pre-qualified contractors without full competition each time.

- Other transaction agreements: Alternative vehicles enabling faster awards, non-traditional participation, and innovative partnerships—valuable for cutting-edge AI.

Vendor Management for AI Success

- Diverse vendor portfolio: Don't depend on single contractors for critical AI. Maintain competitive options.

- Government technical expertise: Build staff AI literacy to evaluate vendor claims and manage performance. Outsourcing all knowledge creates dangerous dependency.

- Modular architecture requirements: Design with modular components and open interfaces. Modularity enables vendor replacement and avoids proprietary lock-in.

- Knowledge transfer mandates: Require documentation, training, and transfer. Your organization must operate AI without being completely dependent on vendors.

- Mission alignment assessment: Seek contractors' understanding of federal missions. Cultural alignment matters as much as technical capability.

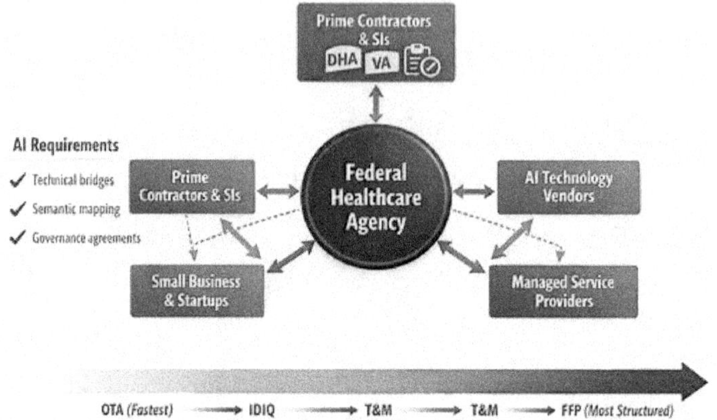

The Clinical-Administrative Duality

Federal healthcare organizations simultaneously operate as clinical delivery systems and massive administrative enterprises. This duality creates fundamentally different AI contexts—different stakeholders, approval processes, risk profiles, and success metrics. Treating all AI the same is one of the most common mistakes.

Clinical AI: High Stakes, High Scrutiny

- Stringent safety requirements: Must meet medical device standards when applicable, undergo clinical validation, and demonstrate safety across populations.

- Clinical governance oversight: Requires approval from clinical committees and medical staff. Clinicians must trust the AI before adoption.

- Liability considerations: Who's responsible when AI provides incorrect recommendations? Clear policies are essential.

- Workflow integration: Must integrate seamlessly without adding burden. Tools that disrupt workflow or create alert fatigue will be circumvented.

- Explainability: Clinicians need to understand reasoning. Black-box systems are inappropriate for clinical decision support.

- Continuous monitoring: Requires ongoing validation for drift, fairness, and safety signals.

Administrative AI: Scale and Efficiency

- Operational efficiency focus: Justified through cost reduction, cycle time improvement, accuracy, and capacity expansion.

- Business process governance: Requires operational leadership approval rather than clinical committees.

- Scalability: Must handle massive transaction volumes and maintain reliability under stress.

- Audit trails: Administrative decisions require documentation for oversight, appeals, and compliance.

- Variable error tolerance: Scheduling optimizers tolerate higher error rates than benefits determination systems.

- Legacy integration: Often must integrate with decades-old systems—technical complexity exceeding AI model complexity.

Managing Across the Duality

- Separate governance pathways: Clinical and administrative AI follow different tracks. Don't force one through the other's process.

- Stakeholder mapping: Identify who must approve, consult, and inform for each domain. Maps differ dramatically.

- Language translation: Frame value in stakeholder terms. Clinicians care about outcomes; administrators care about efficiency.

- Risk calibration: Apply appropriate frameworks. Don't apply clinical safety standards to administrative tools or vice versa.

- Pilot sequencing: Consider administrative AI first to build confidence before tackling high-stakes clinical applications.

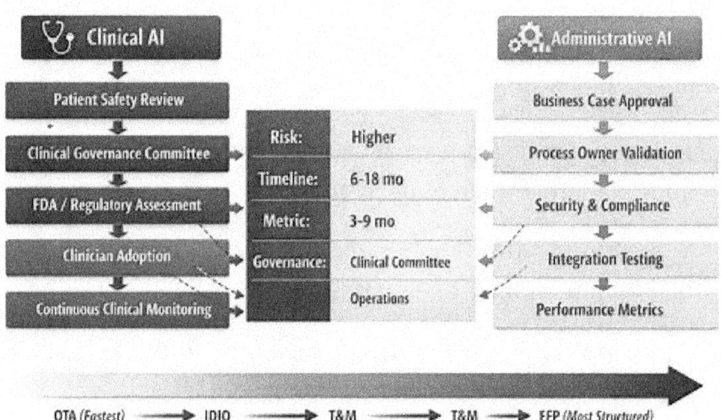

The AI Reality Check

Misconception: "If we just get the right AI tool, we can deploy it across the entire federal healthcare ecosystem."

Reality: There is no "federal healthcare" as a unified deployment target. There are multiple agencies with different EHR platforms, data standards, governance structures, statutory authorities, and risk tolerances. An AI tool that performs brilliantly in VA's VistA environment may fail in DHA's MHS GENESIS. A fraud detection model trained on Medicare claims may be useless for TRICARE claims. A clinical decision support system validated on VA's veteran population may produce biased results for DHA's younger active-duty population.

The 72 AI use cases VA retired in 2025 reinforce this. Many weren't bad technology—they were good technology deployed against shifting infrastructure, fragmented data, or insufficient governance. Plan for the ecosystem you actually operate in, not the unified system you wish existed.

3.13 The Responsible AI Lens

The federal healthcare ecosystem creates specific responsible AI considerations that single-organization deployments don't face:

Cross-agency bias compounding: A model trained on one agency's data may embed population-specific biases that are harmful when applied to another agency's population. VA's older, predominantly male veteran population produces different patterns than DHA's younger, more gender-diverse active-duty population.

Accountability gaps at boundaries: When AI operates across agencies—sharing data between DHA and VA, or CMS and state Medicaid—accountability for errors becomes unclear.

Consent complexity: Beneficiaries may not understand that data flows between agencies or that AI in one agency uses data from another. Transparency is both an ethical obligation and a matter of trust.

Equity across systems: If AI improves care in one agency but not another, the ecosystem creates a two-tier experience for beneficiaries who didn't choose which system they would receive care from.

3.14 What This Means for You on Monday Morning

Understanding the federal healthcare ecosystem isn't academic background—it's operational intelligence shaping every decision you make. When you return to your desk, these realities translate into concrete actions.

Know your agency's position in the ecosystem. Every AI initiative must reflect your agency's unique mission, population, infrastructure, governance, and constraints. For DHA, that's readiness. For VA, it's veteran care and benefits. For CMS, it's beneficiary protection and program integrity. Start with the mission and work backward to the technology.

Map your data reality before committing to AI. Catalog existing systems, data sources, integration capabilities, and technical constraints before selecting tools. Your AI must work with the infrastructure you have, not the infrastructure you wish you had. The 72 retired VA use cases prove that building on shifting ground wastes resources.

Build governance before you build models. Establish approval pathways, risk-reporting mechanisms, and patient-safety coordination before deployment. Forty-five percent of federal AI projects stall at governance gaps. The VA Inspector General's finding that VHA lacks a formal mechanism for tracking clinical AI risks is a warning, not a footnote.

Plan for production from day one. Design pilots with production-grade security, monitoring, and integration built in. When 86 percent of federal AI spending goes to R&D and only 14 percent to deployment, the message is clear: budget for deployment, training, and sustainment from the start.

Identify cross-agency collaboration opportunities. Joint model development, shared platforms, coordinated procurement, and lessons-learned sharing accelerate adoption while reducing duplication.

Distinguish clinical from administrative AI in every conversation: different governance, different risk, different stakeholders, different timelines.

Invest in your contractor strategy. Diverse vendor portfolio. Internal technical expertise. Modular architectures. Knowledge transfer requirements—mission-aligned partners.

3.15   The Ecosystem Navigation Framework

Navigating the federal healthcare ecosystem requires a structured approach. The Ecosystem Navigation Framework provides a repeatable methodology for assessing how ecosystem factors affect any AI initiative. Apply this framework at the start of every project to avoid the most common ecosystem-related failures.

Step One: Mission Alignment Assessment

Before evaluating any AI technology, articulate specifically how the initiative serves your agency's mission. For DHA, quantify readiness impact. For VA, measure improvements in veteran care. For CMS, define beneficiary-protection or program-integrity outcomes. If you can't articulate mission value in one sentence that a non-technical leader would understand, the initiative isn't ready for investment.

Mission alignment isn't a formality. It's your primary defense when budget reviewers question AI spending, oversight bodies ask why resources were allocated, and congressional staff inquire about outcomes. A clearly articulated mission connection turns AI from a technology experiment into a mission capability—and that distinction determines whether initiatives survive budget cycles.

Step Two: Infrastructure Reality Mapping

Map the actual infrastructure your AI must integrate with—not the target architecture, not the vendor's ideal environment, but the systems, data formats, connectivity, and security controls that exist today. Include legacy systems, mid-transition platforms, and planned migrations. Document which facilities have which systems, where data gaps exist, and what integration interfaces are available.

This mapping reveals whether your AI initiative faces a stable deployment environment or a shifting landscape. If your agency is in the mid-EHR transition, the mapping shows exactly which facilities can support deployment today and which require the new platform to complete the migration. If your data sources span multiple agencies, the mapping identifies which data-sharing agreements exist and which require negotiation. Many of VA's 72 retired use cases would have survived if teams had performed rigorous infrastructure mapping before committing to specific technical approaches.

Step Three: Stakeholder and Governance Pathway Mapping

Identify every approval gate between your current state and operational deployment. Map clinical governance committees, technical review boards, privacy offices, security teams, legal counsel, budget authorities, and leadership approval requirements. For each gate, document the specific artifacts required, including risk assessments, security plans, privacy impact assessments, clinical validation reports, and operational readiness reviews.

Build realistic timelines. If your clinical governance committee meets monthly, that's a minimum of 1 month between submission and approval—assuming no revisions. If security authorization requires a full assessment and authorization package, budget 3 to 6 months. Suppose cross-agency data sharing requires a new MOU; budget six to twelve months. Stack these timelines, and you'll understand why federal AI takes longer than commercial deployments—and why front-loading governance preparation accelerates the overall timeline.

Step Four: Contractor and Acquisition Strategy

Determine how you'll acquire AI capabilities. Will you build internally, buy from vendors, or partner through hybrid approaches? Which acquisition vehicles are available—existing IDIQs, OTAs, new competitive procurements? What's the timeline for each vehicle? Which contractors have relevant past performance, security clearances, and domain expertise?

Acquisition strategy isn't a procurement technicality—it determines your implementation timeline. An AI initiative that requires a new competitive procurement may take 12 to 18 months before a contractor is on board. The same initiative using an existing IDIQ task order might begin in 60 to 90 days. Understanding available acquisition vehicles and aligning your strategy accordingly can mean the difference between deploying this fiscal year and deploying three fiscal years from now.

Step Five: Cross-Agency Impact Assessment

Assess whether your AI initiative has cross-agency implications. Will it use data from other agencies? Will its outputs affect beneficiaries served by other systems? Will it create expectations or dependencies that other agencies must support? If cross-agency implications exist, identify the coordination requirements early—data sharing agreements, joint governance structures, aligned standards, and communication protocols.

Cross-agency implications aren't edge cases in federal healthcare. They're common. Veterans receive care from both VA and Medicare providers. Service members transition from DHA to VA. Medicaid beneficiaries interact with both state and federal systems. AI that ignores these boundary crossings will either produce incomplete results or create friction at handoff points. The most effective federal healthcare AI initiatives are designed with ecosystem awareness from inception, not retrofitted when boundary issues surface.

3.16 Manager's Checklist: Ecosystem Readiness

Before launching any AI initiative, confirm you can answer these questions:

- Have I mapped how this initiative connects to my agency's specific mission—not generic "efficiency" but measurable mission outcomes?

- Do I know which systems my AI must integrate with, and are those systems stable or mid-transition?

- Have I identified data sources required, confirmed access, and validated data quality?

- Have I mapped the governance pathway—clinical committee, technical review, privacy, security, legal, leadership—and built approval timelines into my schedule?

- Do I understand whether my AI is clinical, administrative, or both—and applied the appropriate governance track?

- Have I assessed cross-agency implications—does this AI use data from or affect beneficiaries in other agencies?

- Is my contractor strategy aligned—right vehicle, right vendor mix, sufficient internal expertise?

- Have I budgeted for production, not just pilot—deployment, training, monitoring, sustainment?

- Have I incorporated responsible AI from inception—equity testing, explainability, audit trails, privacy?

- Can I explain this to a congressional staffer, an IG auditor, and a veteran beneficiary in language each would understand?

3.17 Lessons from VA's AI Portfolio: A Decision Framework

VA's experience—367 cataloged use cases, 138 in production, 72 retired—provides the most comprehensive dataset on federal healthcare AI deployment outcomes available anywhere in government. These numbers aren't just statistics. They represent real projects, real teams, real budgets, and real consequences for veterans. Analyzing the patterns yields a practical decision framework that any federal healthcare manager can apply.

The tools that succeeded share common characteristics. STORM succeeded because it addressed a measurable mission problem—opioid mortality risk—with a specific, validated intervention pathway: risk scoring integrated into clinical workflows, with mandatory case-review protocols. AI-assisted colonoscopy succeeded because it augmented existing clinical workflows without disruption, produced a directly measurable outcome (adenoma detection rate), and used FDA-cleared technology with established regulatory status. VA GPT succeeded in adoption because it addressed a universal pain point—documentation burden—with immediately visible productivity gains and maintained clear human accountability for all outputs.

The retired tools share different characteristics. They were built on data sources that shifted during EHR modernization. They lacked governance pathways cleared before development began. They duplicated capabilities that other teams were building independently. They were designed for pilot conditions that didn't replicate in production. They assumed infrastructure stability during a period of fundamental infrastructure change.

The pattern is clear: successful federal healthcare AI starts with mission need, validates against real infrastructure, clears governance before building, and plans for production from inception. Failed initiatives start with technology, assume ideal conditions, treat governance as an afterthought, and hope that pilot success translates to production readiness. Every AI initiative you lead should be evaluated against these patterns before committing resources.

This doesn't mean you should avoid ambition. VA's willingness to catalog 367 use cases, deploy 138, and transparently retire 72 reflects healthy organizational learning. The agencies that fail at AI aren't the ones that retire unsuccessful pilots—they're the ones that never pilot at all, or worse, continue funding initiatives that should have been retired years ago. Strategic retirement is a sign of mature portfolio management, not organizational failure. Your goal is to build the institutional infrastructure that enables rapid, rigorous experimentation—and equally rapid, transparent retirement of efforts that don't deliver mission value.

3.18   Chapter Summary

The federal healthcare ecosystem comprises distinct agencies with unique missions, populations, and operating models. Five key takeaways should shape every AI decision you make.

First, the ecosystem is not a system. DHA serves warfighter readiness. VA serves veteran lifetime care. CMS serves beneficiary protection at the population scale. HHS sets policy and oversight. Each operates under different authorities, budgets, and governance. AI strategies must be agency-specific while recognizing cross-agency interdependencies.

Second, data fragmentation is the primary barrier to AI. Technical fragmentation across incompatible platforms, semantic fragmentation across inconsistent terminologies, and organizational fragmentation across agency boundaries compound the difficulty of enterprise AI. Agencies that succeed invest in a data strategy before model development.

Third, the contractor ecosystem shapes AI outcomes. Contract structures determine incentives, acquisition vehicles determine speed, vendor capabilities determine feasibility, and relationship management determines sustainability.

Fourth, clinical and administrative AI require different approaches. Different governance, risk profiles, stakeholders, and timelines—the manager who treats all AI identically either over-governs administrative tools or under-governs clinical ones.

Fifth, real-world evidence—including VA's 72 retired AI use cases and the OIG's clinical AI risk findings—proves that technology alone doesn't determine success. Data readiness, governance infrastructure, production planning, portfolio coordination, and infrastructure alignment determine whether AI delivers mission value or becomes another cautionary tale.

The next chapter examines the policy, compliance, and oversight bodies that govern federal healthcare AI—the rules, frameworks, and requirements that define what's permissible and what's required.

4   Policy, Compliance, and Oversight Bodies

A clinical leader at a VA medical center uses a generative AI tool on her desktop to draft notes, summarize complex charts, and propose differential diagnoses. She has used it in a limited pilot and seen clear productivity gains. Now she wants to expand access across her service line. The problem: the Inspector General has just released a report warning that the VHA lacks a formal mechanism to track risks posed by generative AI in clinical settings. The Office for Civil Rights has reminded agencies of their HIPAA obligations. OMB has released a memorandum restricting the use of certain AI types without documented safeguards. What looked like a straightforward scaling decision now sits at the intersection of policy, compliance, and oversight bodies, each with veto power.

If you manage AI in federal healthcare, this is your reality. Technology decisions are never just technical. They are policy, compliance, and oversight decisions. Executive orders on AI, OMB guidance, NIST frameworks, HIPAA rules, FedRAMP baselines, FISMA requirements, and agency-specific directives all apply—sometimes in overlapping ways, sometimes in conflicting ones.

This chapter gives you a working map of that landscape. It explains who sets the rules, what those rules actually require, how oversight bodies interpret them in practice, and how to build a compliance-by-design mindset that turns policy and oversight from deployment blockers into strategic enablers.

4.1   Why This Matters

In the commercial world, AI leaders often ask, "Can we do this?" In federal healthcare, the better question is, "Should we do this, and under what conditions?" Policy, compliance, and oversight bodies answer that question implicitly every day—through memoranda, audit findings, approval decisions, and enforcement actions. If you do not understand their expectations, you will either move too slowly out of fear or move too fast into risk.

For managers, the goal is not to become a lawyer or policy expert. The goal is to understand the constraints and expectations well enough to shape AI initiatives that are deployable, defendable, and sustainable. That means knowing which levers matter—executive orders, OMB guidance, NIST frameworks, HIPAA and related health laws, FedRAMP and FISMA baselines—and how oversight bodies like OIG, GAO, and Congress look at AI in practice.

4.2   Executive Orders on AI

Presidential executive orders set the tone and direction for how the federal government uses AI. They do not establish every detail, but they define the principles, risk thresholds, and coordination mechanisms that OMB, agencies, and oversight bodies must translate into actionable guidance.

From Innovation to Guardrails

Early federal AI directives focused on encouraging innovation and avoiding over-regulation. More recent executive orders emphasize safety, rights, and accountability, especially for AI systems that affect health, safety, or access to critical services. For federal healthcare managers, this shift means AI is no longer framed primarily as an efficiency opportunity—it is framed as a capability that must be safe, equitable, and rights-preserving by design.

- High-risk system designation: Executive orders increasingly distinguish between low-risk uses (e.g., internal productivity tools) and high-risk systems (e.g., clinical decision support, benefits determinations, law-enforcement-adjacent analytics). High-risk systems trigger additional documentation, testing, and approval requirements.

- Centralized oversight: Orders often direct OMB and specific councils or task forces to coordinate AI policy, leading to cross-government guidance that applies to your agency even if it is not named explicitly.

- Rights and safety focus: Recent directives emphasize safeguarding civil rights, preventing discrimination, ensuring safety, and providing mechanisms for appeal and redress when AI-driven decisions affect individuals.

4.3   What Executive Orders Mean for Managers

You do not implement executive orders directly. You feel them through downstream guidance and reviews. But you should understand three things from them:

- Signal: Executive orders signal what the administration cares about. If the order emphasizes safety and civil rights, expect OMB, OIG, and GAO to ask how your AI initiative addresses those areas.

- Priority: Executive orders often identify priority domains—like healthcare, public benefits, and public safety—where AI must be especially well-governed. Federal healthcare sits squarely in this high-scrutiny zone.

- Direction: Executive orders direct OMB and agencies to create inventories, risk management practices, and reporting mechanisms. Your AI program will be part of those inventories and subject to those practices, whether you plan for it or not.

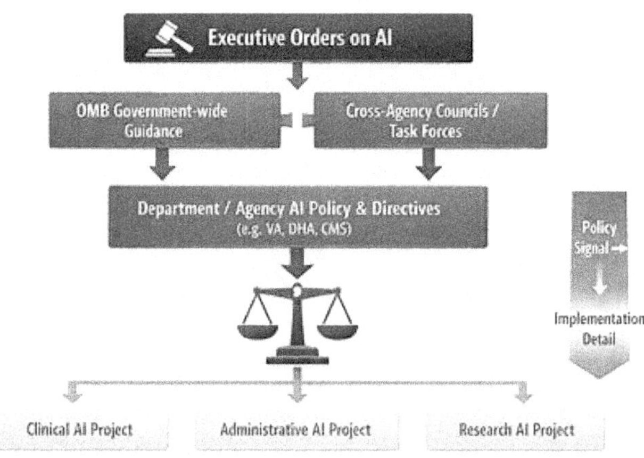

OMB Guidance: The Operational Rulebook

If executive orders set the vision, OMB guidance writes the operational rulebook. OMB memoranda translate presidential direction into concrete requirements for how agencies inventory AI, assess risk, manage vendors, report usage, and obtain approvals. For federal healthcare AI managers, OMB guidance is where abstract principles become specific obligations.

Key Themes in OMB AI Memoranda

AI use case inventories: Agencies must catalog AI use cases, distinguishing between exploratory pilots, production systems, and retired or decommissioned tools.

Risk tiering: OMB guidance pushes agencies to classify AI systems by risk level based on impact on individuals, rights, and safety. High-risk systems require more rigorous governance and documentation.

Human-in-the-loop expectations: OMB emphasizes meaningful human oversight, particularly for systems affecting eligibility, benefits, and health decisions.

Third-party risk management: OMB expects agencies to manage the risks of vendor-provided AI, including documentation, transparency, and alignment with federal standards.

How OMB Guidance Shows Up in Your Work

You will see OMB guidance manifested in three main ways:

Approval checklists: Internal forms and checklists used by your CIO, CISO, or CDO offices will embed OMB expectations.

Portfolio decisions: Leadership will review AI inventories and risk profiles to determine which projects receive funding, which are paused, and which are retired.

Audit and oversight triggers: OIG and GAO reference OMB guidance when assessing whether your agency has adequate controls.

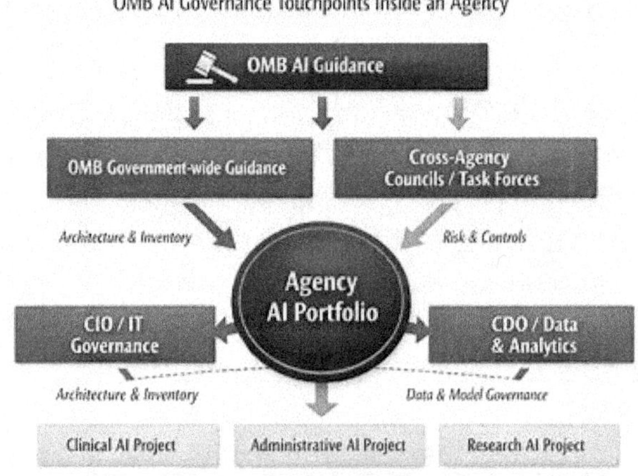

OMB sets expectations; internal governance implements them.

NIST AI Risk Management Framework

The NIST AI Risk Management Framework (AI RMF) is not a law. It is a voluntary framework that has quickly become the de facto standard for how federal agencies talk about AI risk. When OMB, inspectors general, and internal risk committees evaluate AI, they increasingly use the language and structure of the NIST AI RMF.

Core Functions: Govern, Map, Measure, Manage

The AI RMF organizes risk management into four core functions—Govern, Map, Measure, and Manage. For managers, this provides a practical checklist for whether your AI initiative is mature enough to be trusted.

Govern: Establish the structures, policies, roles, and processes to oversee AI.

Map: Understand the context and potential impacts of the AI system—what problem it solves, who is affected, what data is used, and what harms could occur.

Measure: Assess the system's behavior and risks—accuracy, bias and fairness, robustness, and security.

Manage: Act on what you learn—adjust models, update controls, retrain users, or retire systems when risks outweigh benefits.

4.4   Using NIST AI RMF as a Manager, Not a Researcher

You do not need to implement every RMF subcategory to benefit from it. You can use it as a structured way to ask better questions:

Govern questions: Who owns this AI system? Who can approve changes? Who is accountable if something goes wrong?

Map questions: Whose data are we using? Who might be harmed by errors or bias? What decisions will this system influence?

Measure questions: How do we know it works? How does performance differ across subpopulations? How often do we re-evaluate?

Manage questions: What is our process if we detect a problem? Who can pause or turn off the system? How do we communicate changes to users?

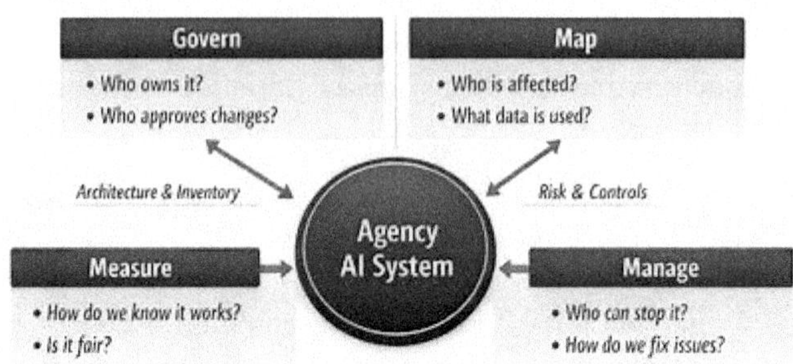

OMB sets expectations; internal governance implements them.

Health Law and Security Baselines: HIPAA, 21st Century Cures, FedRAMP, FISMA

In federal healthcare AI, four pillars recur: HIPAA and related health privacy rules, the 21st Century Cures Act, FedRAMP, and FISMA. Together, they define how data must be protected, how systems must be secured, and how information must flow—or not flow—across systems.

HIPAA and Health Privacy Rules

Any AI system that handles protected health information must comply with HIPAA's privacy and security rules. In practice, that means:

Minimum necessary: AI should only access the data it genuinely needs.

Use and disclosure controls: You must define and enforce who can use AI outputs, for what purposes, and under which authorizations.

Business associate agreements (BAAs): Vendors handling PHI for your AI must sign BAAs with clear security and incident reporting obligations.

21st Century Cures Act and Information Blocking

The 21st Century Cures Act and its information blocking rules push the health system toward more open data exchange. For AI managers, this is both an opportunity and a constraint:

Opportunity: Easier access to standardized data via FHIR APIs and USCDI data classes makes it more feasible to build and integrate AI across EHRs.

Constraint: You cannot design AI workflows that effectively block patient access to their own information or interfere with required information sharing.

4.5   FedRAMP and FISMA: Cloud and System Security

Most modern AI runs on cloud infrastructure. FedRAMP and FISMA define the level of security that infrastructure must meet and how it must be monitored.

- FedRAMP: Standardized security assessment and authorization for cloud services. If your AI uses a cloud platform, you will be asked whether that platform has FedRAMP authorization at the appropriate impact level.

- FISMA: Requires agencies to categorize systems by impact (low, moderate, high), implement corresponding security controls, and maintain ongoing authorization through continuous monitoring.

- Implication: Even if your model is excellent, it cannot go to production on a platform that does not meet FedRAMP/FISMA expectations for your data sensitivity level.

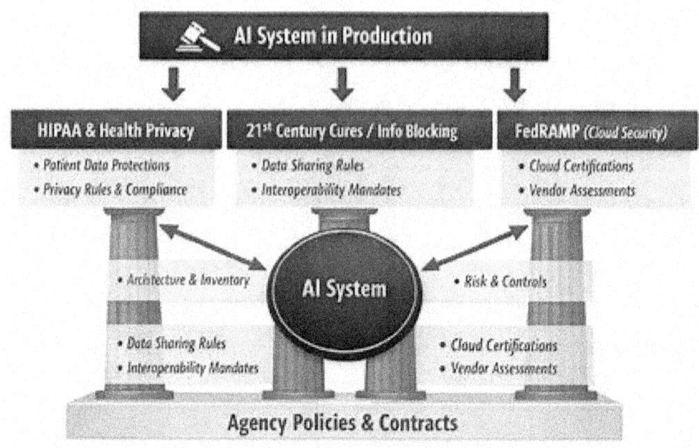

OMB sets expectations; internal governance implements them.

How Oversight Bodies Shape AI Adoption

Oversight in federal healthcare is a continuous condition. Inspectors general, GAO, congressional committees, and watchdog organizations shape how agencies think about AI risks and how aggressively they deploy or pause systems.

4.6 Inspector General (OIG) and GAO

Recent oversight work on AI in VA and other agencies illustrates the pattern:

They look for inventories: Are you tracking where AI is used, for what, and with what safeguards?

They look for governance: Do you have formal mechanisms to identify, track, and resolve AI-related risks—especially in clinical settings?

They look for documentation: Can you show how you validated models, managed bias, trained users, and monitored performance?

They look for response capability: When problems are found, can you pause or adjust AI systems quickly and safely?

Congress and Public Scrutiny

When AI affects veterans, service members, or beneficiaries, congressional interest follows. Hearings, letters, and inquiries often focus on edge cases and failures—not the majority of cases that worked as intended. Your job as a manager is to ensure that when questions arise, you can explain the system's purpose, the safeguards in place, and the steps you took when issues arose.

4.7   The Compliance-by-Design Mindset

The most important shift for federal healthcare AI managers is moving from compliance as a final gate to compliance as a design principle. That means compliance requirements are baked into every phase—not bolted on at the end.

What Compliance-by-Design Looks Like in Practice

- Requirements: For every functional requirement, capture associated compliance requirements (logging, consent, access control, auditability).

- Architecture: Choose data and hosting architectures that already meet FedRAMP/FISMA expectations rather than retrofitting security controls later.

- Vendor selection: Evaluate vendors on compliance posture— HIPAA readiness, FedRAMP status, documentation practices—not just model performance.

- Testing: Treat fairness, explainability, robustness, and security as core test dimensions.

- Operations: Build monitoring, logging, and retraining processes that satisfy both operational needs and oversight expectations.

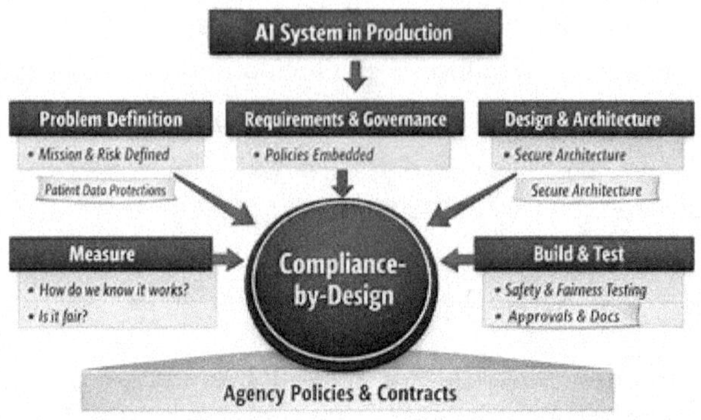

Manager's Checklist: Policy and Compliance Readiness

Before you greenlight an AI initiative, pressure-test it against these questions:

- Can I explain how this initiative aligns with current executive orders on AI (safety, rights, accountability)?

- Do I know which OMB memoranda apply and how my agency has implemented them (inventory, risk tiers, human oversight)?

- Have we mapped this system against the NIST AI RMF functions—Govern, Map, Measure, Manage?

- Have we identified all applicable laws and baselines—HIPAA, 21st Century Cures, FedRAMP, FISMA—and embedded them in requirements?

- Do we have a clear oversight path—who will review, approve, and monitor this system, and how often?

- Can we demonstrate to OIG or GAO how we validated this system and how we will respond if issues emerge?

- Does our vendor or internal team have a compliance track record, not just a strong demo?

- Is compliance treated as a constraint to work within from day one, not a hurdle to clear at the end?

4.8   Chapter Summary

Policy, compliance, and oversight bodies are structural elements of your operating environment. Executive orders set the tone; OMB guidance writes the operational rulebook; NIST provides a shared language for risk; HIPAA and related health laws protect privacy; FedRAMP and FISMA secure infrastructure; and oversight bodies enforce accountability. Ignoring any of these strands guarantees delay at best and program failure at worst.

The managers who succeed with AI in federal healthcare are not those who find ways around policy and oversight; they are those who design within those policies and oversight. They treat compliance as a design input rather than a post hoc review. They use NIST AI RMF to structure their thinking, OMB guidance to shape their portfolios, and oversight findings from other agencies as early warning signals for their own programs. They build AI initiatives that a clinician can use, a beneficiary can trust, a lawyer can defend, an auditor can understand, and a leader can champion.

The next chapter moves from the rulebook to the playbook: specific clinical use cases in federal healthcare where AI already delivers mission-aligned value—and what it takes to make those use cases safe, effective, and scalable in your environment.

5   Clinical Use Cases

An emergency physician at a military treatment facility reviews a patient presenting with fever, elevated heart rate, and altered mental status. The symptoms could indicate sepsis—a life-threatening condition where early treatment dramatically improves outcomes. But they could also indicate a dozen other conditions. The clinician needs to make a decision quickly, but the clinical picture is ambiguous.

In the background, an AI risk-stratification system has been analyzing the patient's vital signs, lab results, medication history, and recent healthcare encounters in real time since the patient arrived. The AI surfaces a high-probability sepsis alert, supported by specific biomarker patterns, vital sign trends, and risk factors drawn from millions of similar cases. The alert doesn't tell the clinician what to do. It tells the clinician what to consider—and provides the clinical context to support rapid, evidence-informed decision-making.

This is clinical AI at its best: not replacing clinical judgment, but augmenting it and not eliminating uncertainty, but reducing it. Not automating care decisions, but accelerating the path from data to insight to action.

This chapter examines clinical use cases in which AI delivers measurable mission value in federal healthcare. It covers clinical decision support, care coordination, imaging and diagnostics, and predictive analytics for risk stratification. For each domain, it explains what works, what does not, and what federal healthcare managers must do to deploy these systems safely and effectively.

5.1   Why This Matters

Clinical AI is where the stakes are highest. Errors in administrative AI may delay payments or frustrate users. Errors in clinical AI can harm patients. That difference shapes everything—approval processes, governance structures, testing requirements, monitoring obligations, and organizational risk tolerance.

For managers, clinical AI requires a fundamentally different mindset than administrative AI. You cannot move fast and break things. You cannot deploy minimum viable products and iterate in production based on user feedback. You must validate thoroughly, deploy carefully, monitor continuously, and maintain explicit human accountability for every clinical decision influenced by AI.

But the mission value justifies the effort. Clinical AI saves lives, prevents complications, reduces suffering, improves clinician efficiency, and enables care that would be impossible without computational support. The managers who succeed with clinical AI are those who combine operational ambition with clinical humility—who push for deployment where evidence supports it while maintaining the discipline to pause when evidence does not.

Clinical Decision Support: AI at the Point of Care

Clinical decision support systems use AI to provide clinicians with knowledge, patient-specific information, and recommendations at the point of care. These systems analyze patient data—clinical history, lab results, vital signs, medications, imaging—and surface insights that inform diagnostic, therapeutic, and monitoring decisions.

5.2 What Defines Effective Clinical Decision Support AI

Not all clinical decision support qualifies as effective. Federal healthcare is littered with alert systems that clinicians ignore, models that produce too many false positives, and tools that disrupt workflow more than they help. Effective clinical decision support AI shares specific characteristics:

- Actionable recommendations: The AI does not just flag risk— it suggests specific evidence-based interventions clinicians can consider.

- Integrated into workflow: The system appears where clinicians work (inside the EHR, not a separate application) and at the moment they need information (during clinical encounters, not hours later).

- Clinically validated: Performance has been demonstrated on real patient populations similar to those the system will serve, with documented accuracy, sensitivity, and specificity.

- Explainable outputs: Clinicians understand why the AI made a recommendation—which data elements drove the assessment, which risk factors were present, and which clinical guidelines apply.

- Respectful of clinical judgment: The system presents recommendations for clinician consideration, not as automated decisions. Clinicians can override recommendations and document their reasoning.

- Alert fatigue mitigation: The system prioritizes high-value alerts and suppresses low-value notifications. Every alert must justify the interruption it creates.

5.3   STORM: Stratification Tool for Opioid Risk Mitigation

VA's Stratification Tool for Opioid Risk Mitigation represents clinical decision support AI done right. STORM analyzes electronic health record data from veterans with active opioid prescriptions or opioid use disorder diagnoses to calculate patient-specific risk scores for overdose and suicide. The system prioritizes patients for interdisciplinary team review and presents clinicians with tailored risk mitigation strategies.

What makes STORM effective is not just the predictive model—it is the entire system design. STORM does not tell clinicians what to do. It tells them which patients need closer attention and provides evidence-based options, such as naloxone co-prescribing, substance use disorder consultation, mental health referral, dose-tapering protocols, or care plan adjustments. Clinicians review the AI-identified cases and apply their judgment about appropriate interventions.

The outcomes speak for themselves. Implementation of STORM is associated with a 22 percent decrease in mortality among high-risk veterans when sites follow required case review protocols. The algorithm did not guarantee that outcome—it was achieved through the combination of accurate risk prediction, clinician review, and evidence-based intervention.

For managers, STORM demonstrates the path to clinical AI success: address a measurable problem (opioid-related mortality), integrate into existing workflows (interdisciplinary team reviews), provide actionable recommendations (specific mitigation strategies), maintain human accountability (clinician-led case review), and measure outcomes (reductions in mortality). STORM did not succeed because it was the most sophisticated model. It succeeded because it was the most operationally sound system.

5.4   Sepsis Prediction and Early Warning Systems

Sepsis is a life-threatening condition where early recognition and treatment dramatically improve survival. But sepsis presents with nonspecific symptoms that overlap with many other conditions, making early identification challenging. AI early warning systems analyze vital signs, lab values, and clinical context in real time to identify patients at elevated risk of sepsis before obvious clinical manifestations appear.

UC San Diego Health deployed an AI tool for sepsis prediction in emergency departments that continuously monitors patient data and alerts care teams when sepsis risk patterns emerge. The system achieved a 17 percent reduction in mortality by enabling earlier antibiotic administration and more aggressive supportive care for high-risk patients. The AI operates silently in the background, analyzing every patient, and surfaces alerts only when risk patterns meet evidence-based thresholds.

The critical design choice: the AI only alerts when it can differentiate sepsis risk from other conditions with high confidence. Systems that alert on every patient with a fever and elevated heart rate create alert fatigue and lose clinician trust. Systems that alert only when specific risk patterns emerge maintain credibility and drive clinical action.

For federal healthcare managers considering AI for sepsis prediction, the key questions are not just about model performance—they are about integration, workflow, and response capacity. Can your EHR infrastructure support real-time data ingestion and scoring? Can your clinical teams respond to AI alerts with appropriate diagnostic and treatment protocols? Do you have the nursing and physician capacity to investigate flagged cases without overwhelming the existing workload? Technology is necessary but not sufficient—operational readiness determines success.

Clinical Decision Support AI Workflow

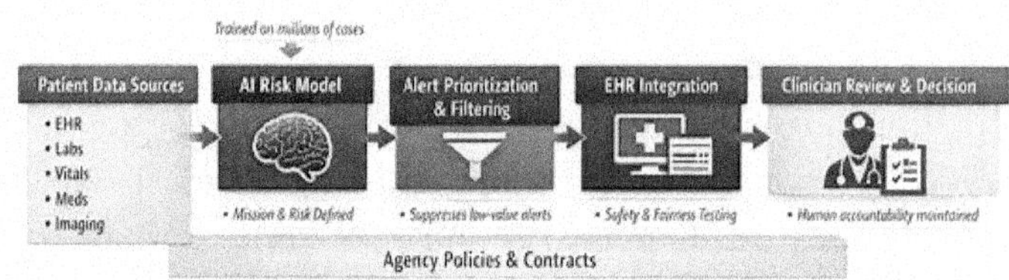

When Clinical Decision Support AI Fails

Not all clinical decision support AI succeeds. Common failure modes provide instructive lessons for managers:

- Alert fatigue: Systems that generate too many alerts—especially false positives—lose clinician trust. Once clinicians ignore alerts, the system becomes ineffective, regardless of model accuracy.

- Workflow disruption: Tools that require clinicians to leave the EHR, log into separate systems, or manually enter data rarely achieve sustained adoption. Integration is not optional.

- Unexplainable recommendations: Black-box systems that provide recommendations without showing supporting evidence frustrate clinicians who need to understand reasoning before acting on suggestions.

- Demographic bias: Models trained on unrepresentative populations may perform poorly or unfairly for patient groups underrepresented in training data—women, racial and ethnic minorities, elderly patients, or those with complex comorbidities.

- Maintenance neglect: AI performance degrades over time as clinical practices evolve, patient populations shift, and data patterns change. Systems deployed without ongoing monitoring and retraining become obsolete.

5.5   Care Coordination: Managing Complexity Across Transitions

Care coordination ensures that patients receive timely, appropriate care across multiple providers, settings, and transitions. For complex patients with multiple chronic conditions, specialist involvement, and frequent healthcare encounters, coordination failures lead to duplicated services, missed follow-ups, medication errors, and preventable complications. AI enhances care coordination by identifying high-risk patients, predicting adverse events, optimizing care pathways, and automating administrative tasks that consume clinician time.

Readmission Risk Prediction and Prevention

Hospital readmissions within 30 days of discharge represent both a quality failure and a financial penalty under federal payment programs. Predictive analytics can identify patients at elevated risk of readmission based on clinical, demographic, and social factors, enabling targeted post-discharge interventions.

Effective readmission risk models incorporate clinical variables—diagnosis, comorbidities, prior admissions, functional status—along with social determinants of health: housing stability, food security, transportation access, and social support networks. Models that rely solely on clinical data typically achieve 65 to 70 percent prediction accuracy. Adding social determinants can push accuracy above 80 percent.

Kaiser Permanente achieved a 12 percent reduction in readmissions by implementing AI predictive analytics across 39 hospitals. The differentiator was not just the model—it was the operational response. High-risk patients identified by the AI received intensive case management, scheduled follow-up appointments before discharge, medication reconciliation, transportation assistance, and care transition coaching. The AI identified who needed help. The operational infrastructure delivered that help.

For federal healthcare managers, readmission-prediction AI is only valuable if you have the care-coordination infrastructure to respond. Can your case management team handle the volume of high-risk patients the AI will identify? Do you have community partnerships to address social determinants? Can you schedule timely post-discharge follow-up appointments? The AI is a targeting mechanism—not a solution on its own.

Chronic Disease Management at Scale

Federal healthcare serves millions of patients with diabetes, hypertension, heart failure, COPD, and other chronic conditions requiring ongoing monitoring and intervention. Managing these populations at scale exceeds human capacity—care teams cannot manually review every patient each week to identify those who are deteriorating or falling out of care.

AI-driven population health platforms continuously analyze patient data to identify individuals whose conditions are worsening, who have missed critical appointments, who require medication adjustments, or who exhibit patterns associated with upcoming exacerbations. These systems generate prioritized worklists for care managers, flagging patients who need outreach, clinical review, or care plan modifications.

The operational model shifts from reactive care (responding when patients call or present to emergency departments) to proactive management (identifying deterioration early and intervening before crises). This requires cultural and workflow changes—care teams must trust AI-generated worklists enough to prioritize them over their own intuitions about which patients need attention.

For managers implementing AI for chronic disease, focus on adoption as much as accuracy. Care teams will not use worklists they do not trust. Start with high-confidence cases where AI predictions align with clinical expectations. Build credibility through demonstrated value. Gradually expand to more nuanced predictions as team confidence grows. Change management is as critical as model development.

Care Transition Support

Transitions between care settings—hospital to home, primary care to specialist, military to veteran healthcare—are high-risk events. Information gets lost, medications change without proper reconciliation, follow-up appointments are not scheduled, and patients receive conflicting instructions from different providers.

AI can smooth transitions by automating information transfer, identifying gaps in care plans, flagging medication discrepancies, and ensuring follow-up appointments are scheduled before discharge. Natural language processing can extract relevant clinical information from discharge summaries and consultation notes, making it accessible to receiving providers without manual chart review.

The joint federal EHR shared by DHA and VA creates unprecedented opportunities for care coordination as service members transition to veteran status. AI systems designed to operate across both organizations can identify veterans at high risk based on their military health history and flag them for proactive VA outreach. This requires cross-agency data sharing agreements, common data standards, and aligned governance—not just technical integration.

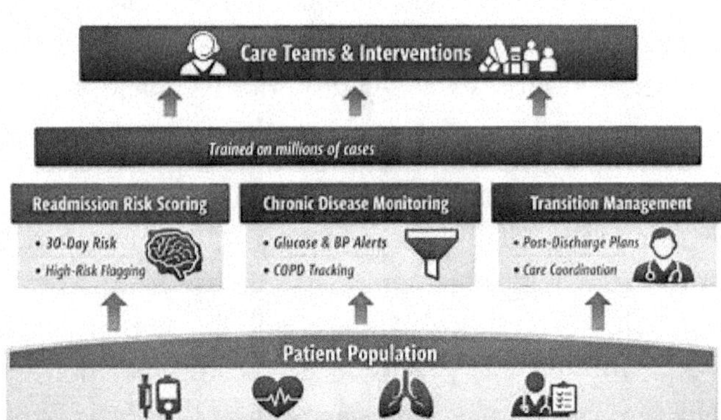

Imaging and Diagnostics: AI as a Second Observer

Medical imaging generates vast volumes of data that require expert interpretation. Radiologists, pathologists, and other imaging specialists review thousands of images annually, looking for subtle abnormalities that indicate disease. AI excels at pattern-recognition tasks when large training datasets are available, making imaging one of the most mature clinical AI domains.

AI-Assisted Colonoscopy: Proven Cancer Prevention

VA's deployment of AI-assisted colonoscopy devices is among the most successful clinical AI implementations in federal healthcare. These FDA-cleared devices use real-time computer vision to highlight potential polyps during colonoscopy procedures, acting as a "second observer" that draws attention to suspicious lesions the endoscopist might otherwise miss.

VA deployed over 115 AI colonoscopy devices across more than 40 facilities starting in late 2022. More than 38,000 procedures have been performed using these devices. Research shows a statistically significant 21 percent increase in the odds of adenoma detection—a direct cancer prevention outcome. VA gastroenterologists already achieved adenoma detection rates of 46 percent, well above the 30 percent benchmark for men in the general population. The AI pushed performance even higher.

What makes AI-assisted colonoscopy particularly manager-friendly is the deployment simplicity. The devices integrate into existing endoscopy workflows with no changes to scheduling, patient preparation, or documentation. Endoscopists do not need extensive training—the AI highlights regions of interest in real time during the procedure. The clinical decision remains entirely with the physician: inspect the highlighted region more closely, biopsy if warranted, or determine it is a false positive and move on.

For managers, AI colonoscopy demonstrates the value of well-defined clinical applications with FDA clearance, published clinical evidence, minimal workflow disruption, and objectively measurable outcomes. This is not experimental AI—it is evidence-based clinical practice that uses AI as an enabling technology.

Chest X-Ray Interpretation and Triage

Chest X-rays are among the most commonly performed imaging studies in healthcare. They screen for pneumonia, pneumothorax, pulmonary edema, lung masses, and dozens of other conditions. AI algorithms trained on millions of X-rays can detect abnormalities with an accuracy exceeding 96 percent for specific conditions such as pneumothorax.

The FDA has cleared multiple AI-powered chest X-ray interpretation systems. These tools analyze portable X-rays in real time and, when they detect life-threatening conditions like collapsed lung, move the image to the top of the radiologist review queue and display alerts to bedside clinicians. This enables faster diagnosis and treatment for time-sensitive conditions.

The operational model preserves radiologist authority. The AI does not replace radiologist interpretation—it prioritizes which images need urgent review and provides preliminary findings that licensed radiologists verify. This triage function is especially valuable in high-volume emergency departments and intensive care units where imaging volumes can overwhelm reading capacity during surge conditions.

For federal healthcare managers, chest X-ray AI presents procurement and integration challenges. FDA clearance is a starting point, not an endpoint. You must validate that the AI performs accurately with your X-ray equipment in your clinical settings for your patient population. Models trained primarily on academic medical center data may perform differently when evaluated on rural community hospital demographics and equipment. Vendor claims must be validated locally before full deployment.

Radiology Workflow Optimization

Beyond individual image interpretation, AI optimizes radiology workflows by automating image preprocessing, identifying urgent findings for prioritization, extracting structured data from reports, and routing studies to subspecialists with relevant expertise. These workflow optimizations reduce radiologists' cognitive load, decrease report turnaround time, and improve the appropriateness of studies.

AI can analyze radiology orders to identify studies unlikely to yield clinically useful information based on ordering indication, clinical history, and prior imaging. This clinical decision support at the ordering stage reduces unnecessary imaging, decreases radiation exposure, and frees imaging capacity for higher-value studies.

Pathology and Histopathology Analysis

Digital pathology combined with AI enables automated analysis of tissue samples, cell morphology, and biomarker expression. AI can quantify tumor characteristics, identify metastatic disease in lymph nodes, grade cancer aggressiveness, and predict treatment response based on histological patterns.

For federal healthcare, digital pathology AI faces infrastructure barriers. Many pathology departments still rely on glass slides and optical microscopes rather than digitized whole-slide imaging systems. Deploying AI requires first digitizing pathology workflows—a significant capital investment that must be justified on broader operational grounds, not just AI capability.

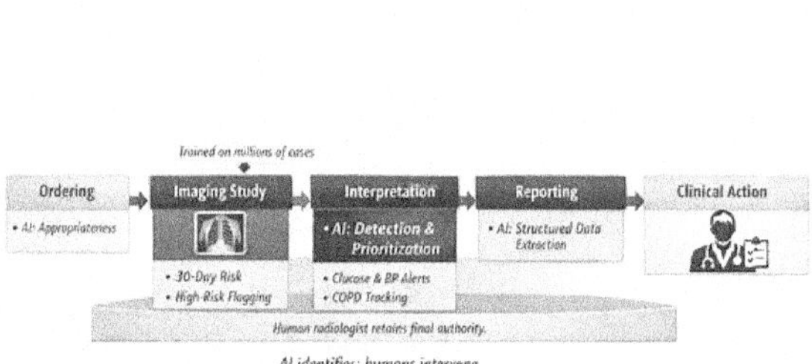

When Imaging AI Underperforms

Imaging AI failures provide critical lessons:

Training data mismatch: Models trained on one population or imaging equipment type may perform poorly when deployed on different populations or equipment. A model trained on urban academic medical center chest X-rays may not generalize to rural facility chest X-rays taken with older equipment.

Rare-condition challenges: AI trained on common conditions may fail to detect rare diseases absent from or underrepresented in the training data. Radiologists catch these rare conditions through experience and vigilance; AI may not.

Image quality sensitivity: AI performance degrades more rapidly with poor image quality than human expert performance. Artifacts, motion blur, positioning errors, and equipment limitations all affect AI accuracy.

Over-reliance risk: When AI is highly accurate, clinicians may unconsciously defer to AI judgments rather than independently evaluating images. This creates risk when AI fails on edge cases.

False positive burden: AI systems with high sensitivity but low specificity flag many findings that turn out to be clinically insignificant, increasing radiologist workload rather than reducing it.

Predictive Analytics for Risk Stratification: From Hindsight to Foresight

Traditional healthcare analytics describe what happened: how many patients were admitted, what treatments were provided, and what outcomes occurred. Predictive analytics describe what is likely to happen: which patients will develop complications, who will require intensive interventions, and where capacity constraints will emerge. This shift from retrospective reporting to prospective forecasting changes how managers allocate resources, target interventions, and plan operations.

What Risk Stratification AI Actually Predicts

Risk-stratification AI does not predict individual patients' futures with certainty—it calculates probabilities based on patterns observed in historical data. A 30 percent readmission risk score indicates the patient resembles a population in which 30 percent were readmitted within 30 days. It does not mean this specific patient has a 30 percent chance of readmission.

This distinction matters for clinical communication and intervention design. High-risk scores identify patients who warrant closer attention and proactive intervention—not patients destined for poor outcomes regardless of what clinicians do. The goal of risk stratification is to enable interventions that change outcomes, not to predict unchangeable futures passively.

Multi-Condition Risk Models

The most powerful risk stratification systems address multiple conditions simultaneously. Rather than separate models for heart failure exacerbation, diabetic complications, fall risk, and medication non-adherence, integrated platforms calculate risk across multiple dimensions and present clinicians with unified patient risk profiles.

This integrated approach reflects clinical reality. Patients do not have isolated conditions—they have interacting comorbidities where heart failure affects diabetes management, cognitive decline impacts medication adherence, and mobility limitations increase fall risk. Siloed risk models miss these interactions. Comprehensive models capture them.

For managers, integrated risk platforms require comprehensive data access, including lab results, vital signs, medications, diagnoses, procedures, imaging, clinical notes, and social determinants. Data fragmentation is the primary barrier to effective risk stratification. If your organization cannot aggregate patient data from multiple systems into a unified analytic environment, your risk-stratification AI will be limited, regardless of model sophistication.

Population Health Segmentation

Not all patients need the same level of clinical attention. Population health segmentation uses AI to categorize patients into risk tiers—low, moderate, high, or very high—based on predicted resource needs and the probability of adverse events. This segmentation drives differentiated care models, with very high-risk patients receiving intensive case management and low-risk patients receiving standard care with automated monitoring.

Effective segmentation requires operational follow-through. Identifying high-risk patients is pointless if care delivery remains one-size-fits-all. Managers must design differentiated care pathways: enhanced primary care for moderate-risk patients, care coordination for high-risk patients, and intensive interdisciplinary management for very high-risk patients. The AI enables the targeting. The care model delivers the value.

5.6   Predictive Models for Capacity Planning

Beyond individual patient risk, AI forecasts aggregate demand: expected admission volumes, ICU bed needs, surgical case mix, clinic visit patterns, and specialty referral rates. These forecasts enable proactive capacity management—adjusting staffing, managing elective scheduling, and coordinating across facilities to match supply with predicted demand.

Federal healthcare systems with multiple facilities can use predictive demand models to load-balance across sites, directing patients to facilities with available capacity and avoiding bottlenecks. This requires real-time data sharing, coordinated scheduling systems, and governance structures that prioritize system-level optimization over individual facility metrics.

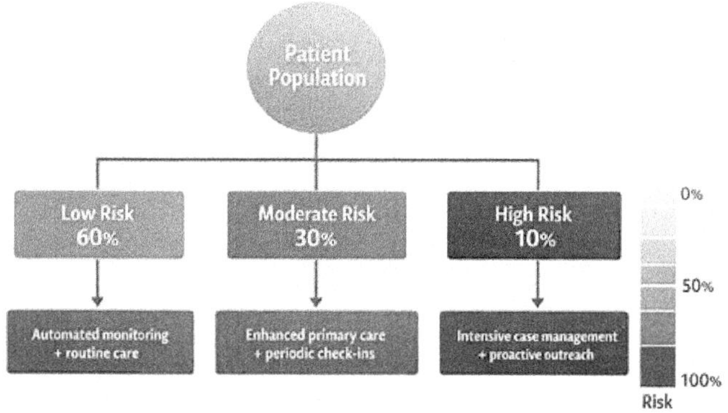

AI segments; care model adapts.

Temporal Dynamics and Model Drift

Risk stratification models degrade over time. Patient populations change, clinical practices evolve, disease epidemiology shifts, and data collection processes are modified. A model trained on 2022 data may perform poorly in 2026 because the patterns it learned no longer apply.

Model drift requires continuous monitoring and periodic retraining. Managers must establish operational processes to track model performance over time—not just at deployment but monthly or quarterly thereafter. Key performance indicators include discrimination (does the model still separate high-risk from low-risk patients effectively?), calibration (do predicted risk percentages match observed event rates?), and fairness (does performance remain consistent across demographic groups?).

When performance degrades, models must be retrained on recent data, recalibrated, and revalidated before continued use. This is not a one-time deployment project—it is an ongoing operational capability. Budget for it. Staff for it. Govern it.

5.7 Avoiding Risk Score Misuse

Risk scores can be misused in ways that harm patients:

- Rationing care: Using risk scores to deny services rather than target interventions. A high mortality risk score should trigger intensive support, not a decision to withhold treatment.

- Self-fulfilling prophecies: When clinicians see high-risk scores, they may unconsciously reduce treatment intensity ("this patient will likely die anyway") rather than intensify it.

- Discrimination: Risk models that disadvantage specific demographic groups—whether through biased training data, proxy variables, or systemic healthcare disparities encoded in historical patterns.

- Gaming incentives: When payment or performance metrics depend on risk-adjusted outcomes, organizations have incentives to inflate risk scores to make actual outcomes look better by comparison.

- Ignoring social determinants: Clinical risk models that ignore social factors may assign high-risk scores to patients whose primary needs are housing, food security, and transportation rather than medical interventions.

Managers must establish governance policies that define acceptable and unacceptable uses of risk scores, with mechanisms to audit actual use and intervene when misuse is detected.

- Guardrails for Clinical AI: The Safety Infrastructure That Makes Deployment Possible

Clinical AI without guardrails is clinical negligence. The question is never whether to implement guardrails—it is which guardrails, at what thresholds, and with what oversight. This section describes the safety infrastructure federal healthcare managers must build before deploying clinical AI.

5.8 Human-in-the-Loop Requirements

Clinical AI should support human decisions, not replace them. Human-in-the-loop design ensures that AI outputs inform clinical judgment rather than automate clinical decisions. The clinician reviews AI recommendations, evaluates supporting evidence, applies contextual knowledge the AI lacks, and makes the final decision with full accountability.

Different clinical contexts require different levels of human oversight:

- Human-on-the-loop: AI makes recommendations that clinicians can accept, modify, or override. Examples include clinical decision support alerts, risk scores, and care pathway suggestions. Clinicians retain full authority but benefit from AI insights.

- Human-in-command: AI provides information or analysis but makes no recommendations. Examples include imaging detection results displayed for radiologist interpretation or summarized patient histories for clinician review. The AI surfaces data; humans decide what it means and what to do with it.

- Human veto authority: AI performs tasks autonomously, but clinicians can review and reverse decisions. This model is rare in federal healthcare due to liability and safety concerns. Even when technically feasible, human review before action is typically required.

For federal healthcare managers, human-in-the-loop is not just about interface design—it is about accountability structures. When an AI-influenced decision leads to patient harm, who is responsible? The answer must be clear: the clinician who made or approved the decision. AI is a tool the clinician uses, not an independent actor. This legal and ethical reality shapes every design choice.

5.9   Explainability and Transparency Standards

Clinicians must understand why AI systems make recommendations. Black-box models that produce risk scores without explanation are clinically unacceptable. Effective clinical AI provides:

- Feature importance: Which data elements most influenced the AI output? For a high readmission risk score, were the primary drivers prior admissions, medication complexity, social isolation, or functional decline?

- Cohort comparison: How does this patient compare to the population the model was trained on? Is this patient within the model's validated range or an edge case where predictions may be unreliable?

- Clinical guidelines linkage: Which evidence-based guidelines support the AI recommendation? If the AI suggests sepsis screening, which specific sepsis criteria and clinical decision rules apply?

- Uncertainty quantification: How confident is the model in its prediction? A 70 percent risk score with wide confidence intervals should be interpreted differently from a 70 percent score with narrow intervals.

- Temporal context: How has the patient's risk changed over time? Is this a sudden spike requiring immediate attention or a gradual increase that warrants scheduled follow-up?

Explainability serves clinical, legal, and organizational purposes. Clinically, it enables appropriate interpretation. Legally, it supports documentation of decision-making rationale. Organizationally, it builds trust, encouraging appropriate AI use rather than blanket rejection or uncritical acceptance.

5.10   Bias Detection and Mitigation

Clinical AI can perpetuate or amplify healthcare disparities if models perform differently for different demographic groups. Bias can enter through the composition of training data, feature selection, outcome definition, or optimization objectives. Managers must establish processes to detect and mitigate bias before deployment and monitor for bias emergence during operation.

Pre-deployment bias assessment includes:

Demographic performance analysis: Test model accuracy, sensitivity, specificity, and calibration separately for relevant demographic groups—race, ethnicity, gender, age, rural versus urban, primary language, socioeconomic status.

Feature fairness review: Identify features that may serve as proxies for protected characteristics. ZIP code, for example, often correlates with race and socioeconomic status.

Outcome definition scrutiny: Ensure that outcome labels reflect clinical reality rather than healthcare access disparities. Using "did not receive follow-up care" as a negative outcome may penalize patients who lacked transportation rather than those with poor health.

Training data representativeness: Compare training data demographics to the population the AI will serve. Underrepresentation signals risk of poor performance.

Post-deployment bias monitoring tracks whether AI recommendations lead to differential care patterns by demographic group and whether outcomes differ by group after controlling for clinical severity.

Clinical AI Safety Layers

Each layer reinforces the others.

Alert Threshold Calibration

AI systems that generate alerts must balance sensitivity (catching all relevant cases) with specificity (avoiding false positives). Too sensitive, and clinicians drown in alerts. Too specific, and the AI misses critical cases. Threshold calibration is a clinical and operational judgment, not just a statistical optimization.

Effective threshold calibration requires:

- Clinical input: Clinicians who will use the system define acceptable false positive and false negative rates based on clinical consequences and workload capacity.

- Pilot testing: Deploy with conservative thresholds (higher specificity, fewer alerts) to build trust, then gradually adjust based on clinician feedback and outcome data.

- Context awareness: Different thresholds for different care settings. ICU alerts may tolerate higher false positive rates than outpatient clinic alerts because ICU clinicians expect more monitoring and intervention.

- Feedback loops: Clinicians should be able to flag inappropriate alerts, creating data for threshold refinement and model improvement.

5.11 Human-in-the-Loop Requirements: Designing for Appropriate Reliance

Human-in-the-loop design is not a checkbox—it is a design philosophy that shapes every interaction between clinician and AI. The goal is appropriate reliance: clinicians use AI when it adds value, override it when clinical judgment dictates, and maintain full situational awareness rather than passively deferring to AI outputs.

The Spectrum of Automation

Not all clinical AI requires the same level of human involvement. The appropriate level depends on task risk, uncertainty, and consequence of error:

Recommendation with required review: AI suggests an action; clinician must explicitly accept or reject with documentation. Examples: medication dosing recommendations, care pathway suggestions.

Alert with response expectation: AI flags a condition; clinician must acknowledge and either act or document why no action is needed. Examples: sepsis alerts, fall risk notifications.

Information display without action expectation: AI provides data or analysis; clinician decides whether and how to use it. Examples: risk scores on patient summary screens, imaging detection results.

Automated action with notification: AI takes action automatically but notifies clinician who can review and reverse. Rare in federal clinical settings. Examples might include automated vital sign monitoring with escalation protocols.

Fully autonomous action: AI acts without human involvement. Essentially prohibited in federal healthcare for clinical decisions due to accountability and liability requirements.

For most clinical AI in federal healthcare, the appropriate level is recommendation with required review or alert with response expectation. This preserves clinical accountability while enabling AI value.

Interface Design for Appropriate Reliance

The interface between clinician and AI shapes reliance patterns. Poor interface design encourages either over-reliance (accepting all AI recommendations without review) or under-reliance (ignoring all AI recommendations regardless of merit). Good interface design supports appropriate reliance.

Design principles for appropriate reliance:

Show confidence: Display AI confidence in predictions alongside the predictions themselves. Clinicians should know when AI is certain versus uncertain.

Highlight uncertainty: Flag edge cases where the patient differs significantly from the training population or where conflicting data elements create ambiguous predictions.

Make disagreement easy: Overriding AI recommendations should require minimal clicks and no justification beyond standard clinical documentation. Do not make disagreement bureaucratically burdensome.

Provide context: Show which data elements drove the AI recommendation and how the patient compares to similar cases. Enable clinicians to verify AI reasoning.

Avoid premature commitment: Do not auto-populate order sets or treatment plans based on AI recommendations. Require explicit clinician selection to prevent mindless acceptance.

Support learning: When clinicians override AI, offer optional feedback mechanisms that capture reasoning. This improves AI over time and demonstrates responsiveness to clinical input.

Training Clinicians to Use AI Appropriately

Deploying AI without training clinicians is deployment malpractice. Effective training covers not just how to use the interface but how to interpret outputs, when to trust recommendations, when to be skeptical, and what the AI can and cannot do.

AI training for clinicians should include:

What the AI was trained on: Population characteristics, data sources, time periods. This helps clinicians assess whether their patients resemble the training population.

What the AI predicts well: Conditions or scenarios where accuracy is high and recommendations are reliable.

What the AI predicts poorly: Known limitations, edge cases, demographic groups with lower performance.

How to interpret outputs: What risk scores mean, how confidence intervals work, which features drive predictions.

When to override: Clinical scenarios where AI may be systematically wrong or where clinical context overrides statistical patterns.

How to report problems: Mechanisms to flag incorrect predictions, inappropriate alerts, or system failures.

Training should not be one-time onboarding—it should be ongoing as AI systems evolve, new use cases emerge, and organizational experience accumulates.

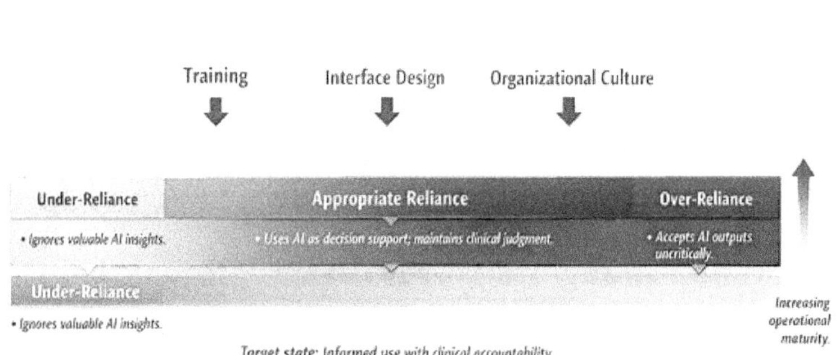

Accountability Frameworks

When clinical AI influences a decision that results in patient harm, accountability must be clear. Ambiguity about who is responsible creates legal risk, chills adoption, and undermines patient trust. Federal healthcare requires explicit accountability frameworks established before AI deployment.

Key accountability principles:

Clinician accountability: The clinician who makes or approves an AI-influenced decision is accountable for that decision. AI is a tool; clinicians are decision-makers.

Organizational responsibility: Healthcare organizations are responsible for selecting appropriate AI tools, validating their performance, training clinicians properly, and monitoring for safety issues.

Vendor obligations: AI vendors are responsible for accurate performance claims, transparent documentation, and prompt notification of newly discovered safety issues or performance limitations.

Regulatory oversight: FDA clearance establishes baseline safety and effectiveness but does not eliminate organizational validation obligations or clinician accountability.

Documentation requirements: Clinicians must document AI-influenced decisions with sufficient detail to demonstrate appropriate use of AI outputs in clinical reasoning.

These accountability principles must be incorporated into medical staff bylaws, credentialing requirements, quality assurance programs, and risk management policies before clinical AI deployment.

5.12   The Manager's Checklist: Deploying Clinical AI Responsibly

Before deploying clinical AI in federal healthcare, managers must ensure the following conditions are met:

- **Clinical Validation and Evidence**

- AI performance has been validated on patient populations similar to yours.

- Accuracy, sensitivity, specificity, and other performance metrics meet clinical standards for the intended use.

- Bias testing has been performed across relevant demographic groups.

- Clinical evidence of benefit (improved outcomes, reduced errors, enhanced efficiency) exists and is documented.

- FDA clearance or authorization exists if the AI qualifies as a medical device.

- **Integration and Workflow**

- AI integrates into existing clinical workflows without requiring separate logins, applications, or manual data entry.

- Alerts and recommendations appear at the point of care when clinicians need them.

- The system does not disrupt clinical efficiency or create alert fatigue.

- Workflow changes required for AI adoption are documented and clinician-tested.

- Technical infrastructure (EHR connectivity, data pipelines, compute resources) is in place.

- **Governance and Safety**

- Accountability frameworks are established and documented.

- Human-in-the-loop requirements are defined and enforced.

- Explainability standards are met—clinicians understand why AI makes recommendations.

- Incident reporting mechanisms exist for AI-related safety issues.

- Monitoring processes track AI performance, bias, and drift over time.

- Governance committee oversight includes clinical leadership with AI accountability.

- **Training and Change Management**

- Clinicians receive training on AI capabilities, limitations, appropriate use, and override procedures.

- Training materials address specific clinical scenarios and edge cases.

- Ongoing education is planned as AI systems evolve.

- Champions are identified to support peer adoption and address concerns.

- Feedback mechanisms capture clinician experience and inform system improvements.

- **Operational Readiness**

- Care coordination infrastructure exists to respond to AI-identified high-risk patients.

- Staffing levels are sufficient to act on AI-generated worklists and alerts.

- Response protocols are defined for different alert types and risk levels.

- Cross-functional teams (clinical, IT, quality, legal, risk management) are aligned.

- Budget includes not just AI licensing but also implementation, training, monitoring, and maintenance costs.

- **Legal and Compliance**

- Privacy impact assessments are completed.

- Data use agreements cover AI training and operation.

- Informed consent processes address AI use where required.

- Liability insurance covers AI-augmented clinical decisions.

- Documentation standards meet regulatory and accreditation requirements.

- Medical staff bylaws and credentialing policies reflect AI use.

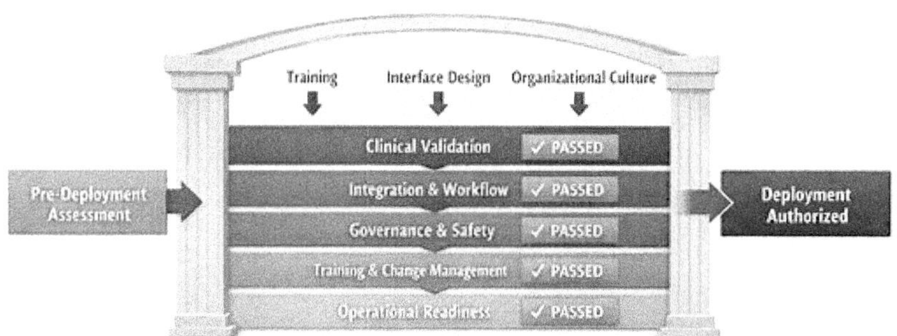

All gates must be satisfied before deployment.

5.13 Chapter Summary

Clinical AI represents the highest-stakes application of AI in federal healthcare. The decisions these systems inform directly affect patient safety, quality of care, and clinical outcomes. This chapter examined four major clinical AI domains—clinical decision support, care coordination, imaging and diagnostics, and predictive analytics—and established the guardrails necessary for safe, effective deployment.

5.14 Key Takeaways

- Clinical decision support AI is effective when it provides actionable recommendations, integrates into workflow, maintains explainability, and respects clinical judgment. VA's STORM system demonstrates the model: accurate risk prediction, clinical review, and evidence-based intervention equal measurable mortality reduction.

- Care coordination AI identifies high-risk patients and optimizes transitions, but the value depends on operational capacity to respond. Predictive models are targeting mechanisms, not solutions. Readmission risk scores are meaningless without a case management infrastructure to intervene.

- Imaging AI has matured rapidly, with FDA-cleared systems demonstrating measurable clinical value. VA's AI-assisted colonoscopy deployment achieved a 21 percent increase in adenoma detection—directly preventing cancer. But imaging AI requires local validation; models must perform accurately on your equipment, your patient population, and your clinical context.

- Predictive analytics for risk stratification shifts healthcare from reactive to proactive. Models forecast complications, deterioration, and resource needs, enabling early intervention. But models degrade over time through drift; continuous monitoring and retraining are operational requirements, not optional enhancements.

- Guardrails are not obstacles to clinical AI—they are the infrastructure that makes deployment possible. Human-in-the-loop design, explainability standards, bias monitoring, and accountability frameworks protect patients, clinicians, and organizations. Clinical AI without guardrails is clinical negligence.

- Human-in-the-loop requirements define the relationship between clinician and AI. The goal is appropriate reliance: clinicians use AI insights when valuable, override when clinical judgment dictates, and maintain full accountability for decisions. Interface design, training, and organizational culture shape whether reliance patterns are appropriate.

- Federal healthcare managers deploying clinical AI must validate performance, integrate into workflow, establish governance, train clinicians, ensure operational readiness, and address legal requirements before deployment. The pre-deployment checklist is not bureaucratic overhead—it is the minimum standard of responsibility.

5.15   What to Do on Monday Morning

- If you are considering clinical AI deployment, start with the pre-deployment checklist. Assess your organization's standing on each dimension. Address gaps before procurement.

- If you have clinical AI in operation, audit compliance with guardrails. Are human-in-the-loop requirements enforced? Is performance monitored? Do incident reporting mechanisms exist and function? Address deficiencies immediately.

- If you experienced an AI-related safety issue, activate your incident response process. Report through established channels, document thoroughly, investigate root causes, and implement corrective actions. Do not minimize or rationalize.

- If clinicians are not using deployed AI, investigate why. Is it alert fatigue? Workflow disruption? Lack of trust? Poor explainability? Address root causes rather than mandating use.

- If you are building business cases for clinical AI, emphasize validated clinical outcomes over technology sophistication. Leadership responds to mortality reduction, complication prevention, and quality improvement—not model accuracy or computational elegance.

Clinical AI done right saves lives, prevents complications, and enables care that would be impossible without computational support. Clinical AI done wrong harms patients, exposes organizations to liability, and undermines trust in technology. The difference is not the sophistication of the algorithms—it is the discipline of the implementation.

Managers who succeed with clinical AI are those who combine operational ambition with clinical humility, who push for deployment where evidence supports it while maintaining the discipline to pause when evidence does not, and who build the governance infrastructure to detect and correct problems before they harm patients. This is not innovation theater—it is mission execution where the stakes are life and death.

6 Administrative and Operational Use Cases

A veteran submits a disability compensation claim for multiple service-connected conditions. The claim includes a 500-page service medical record, five independent medical opinions, deployment documentation from three assignments, and evidence for seven separate conditions. In the legacy process, a claims processor would spend days reading through documents, manually extracting relevant information, cross-referencing medical evidence against rating criteria, and drafting a decision memo from scratch.

Today, that same claim enters VA's Automated Decision Support system. Within minutes, AI extracts relevant medical findings, maps them to specific conditions, identifies which pages contain the required evidence for each condition, and prepopulates draft decision language based on rating schedule criteria. The claims processor reviews the AI-generated summary, verifies the interpretation of the evidence, applies judgment to ambiguous cases, and finalizes the decision. What once took 27 days now takes 12 hours. The processor handles nine times more claims without sacrificing accuracy.

This is administrative AI delivering measurable mission value: faster benefits delivery to veterans, reduced backlog, improved processor productivity, and better resource utilization—all while maintaining human accountability for every decision. The AI did not replace the claims processor. It eliminated the tedious, time-consuming work, allowing the processor to focus on judgment, verification, and decision-making.

This chapter examines administrative and operational use cases in which AI improves federal healthcare efficiency without directly affecting clinical care. It covers claims processing automation, scheduling optimization, contact center augmentation, supply chain and logistics, revenue cycle management, and fraud detection. For each domain, it explains what works, what managers must do to deploy successfully, and how to measure value.

6.1   Why This Matters

Administrative and operational AI may lack the drama of clinical applications, but the mission impact is substantial. Every day, claims processing delays result in delayed benefits for veterans. Every missed appointment wastes clinical capacity that could be used to serve other patients. Every supply stockout risks canceled surgeries. Every dollar lost to fraud is a dollar unavailable for patient care. Administrative inefficiency directly undermines mission delivery.

For managers, administrative AI offers advantages that clinical AI does not: lower risk, faster deployment, clearer ROI, and fewer regulatory barriers. You can pilot administrative AI without FDA clearance, clinical validation, or patient safety reviews. You can measure success through throughput, cost reduction, and cycle time—not health outcomes. You can iterate quickly based on operational metrics rather than waiting for long-term clinical evidence.

But administrative AI still requires discipline. Models trained on biased historical data can perpetuate discriminatory patterns in benefits decisions. Poorly designed automation can frustrate users and reduce service quality. Systems that eliminate human review can make errors at scale without detection. The stakes may be lower than clinical AI, but the accountability requirements remain.

6.2   Claims Processing: From Months to Hours Through Intelligent Automation

Claims processing in federal healthcare involves verifying eligibility, reviewing supporting documentation, applying complex rules and regulations, calculating payment or benefit levels, and communicating decisions. Historically, this process was entirely manual—claims processors read every document, extracted relevant facts, looked up applicable rules, and drafted decisions by hand. AI transforms this workflow by automating information extraction, pre-populating decision templates, and surfacing the specific evidence processors need to review.

VA Claims Processing: The National Model

The Department of Veterans Affairs processes over three million disability compensation and pension claims annually. In 2025, VA achieved record productivity while simultaneously reducing its backlog by 57 percent. This outcome was not achieved by hiring more claims processors—it was achieved through intelligent automation that multiplied the effectiveness of processors.

VA deployed multiple AI systems across the claims lifecycle:

Automated Decision Support (ADS): Machine learning extracts relevant information from medical records, maps findings to claimed conditions, and prepopulates portions of decision memos. The system does not make claims decisions—it retrieves and organizes the information processors need to make decisions.

Acceptable Clinical Evidence (ACE) optimization: AI identifies when existing medical evidence is sufficient to decide a claim without ordering additional examinations. This eliminates unnecessary in-person exams, reduces processing time, and decreases costs without compromising decision accuracy.

Document classification and routing: Natural language processing automatically categorizes incoming documents, routes them to appropriate work queues, and flags urgent cases requiring priority handling.

Pre-adjudication quality review: AI scans draft decisions before finalization to identify missing evidence, inconsistent reasoning, or potential rating errors. Processors can correct issues before decisions are finalized rather than through costly appeals.

The results demonstrate what well-implemented administrative AI achieves: processing time dropped from 27 days to 12 hours for cases where ADS applies; processor accuracy improved through automated quality checks; backlog decreased by over 57 percent; and veterans received life-changing benefits faster. VA explicitly designed these systems to augment human decision-making, not replace it. Every claim decision still requires human review and approval. AI handles the information retrieval and organization; humans apply judgment and make decisions.

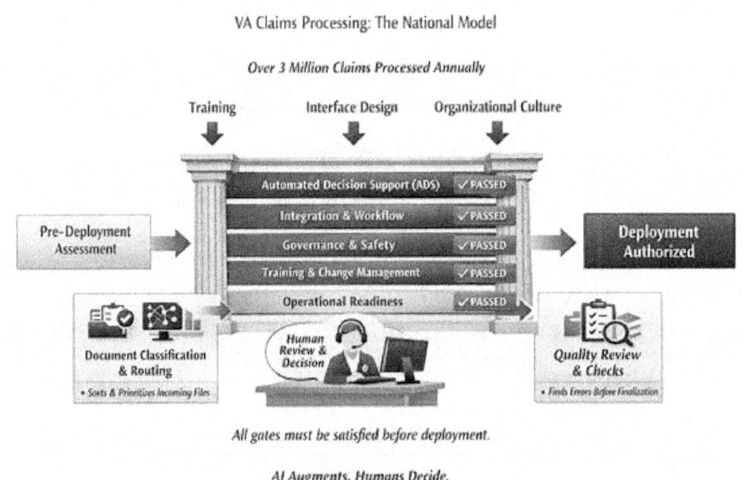

The Anatomy of Effective Claims Automation

Not all claims automation succeeds. Systems that attempt to fully automate decision-making without human review face challenges related to accuracy, fairness, and accountability. Systems that provide minimal assistance waste the AI investment. Effective claims automation follows specific design principles:

Augmentation, not replacement: AI handles high-volume, low-judgment tasks (document extraction, evidence retrieval, template population). Humans handle complex interpretation, ambiguous cases, and final decision authority.

Explainable recommendations: When AI prepopulates decision language or suggests rating levels, it shows which evidence supports the recommendation and which rules or criteria apply. Processors can verify AI reasoning before accepting recommendations.

Progressive automation: Start with straightforward cases where the evidence is clear and the rules are unambiguous. As confidence grows and systems improve, gradually expand to more complex cases. Never automate the hardest cases first.

Human override capability: Processors can reject or modify AI recommendations at any point without technical barriers or administrative burden. Disagreement with AI should be as easy as normal work.

Feedback loops: When processors override AI recommendations, capture the reasoning behind them. Use this data to improve model accuracy and identify systematic errors.

Fairness monitoring: Track whether AI recommendations differ systematically by demographic groups. Claims decisions based on biased historical patterns can perpetuate discrimination even when current processors are unbiased.

Beyond Veterans Affairs: Claims Automation Across Federal Healthcare

While VA has achieved the most visible claims automation success, the model applies across federal healthcare programs:

Medicare and Medicaid claims processing: CMS processes over one billion claims annually. AI automates prior authorization reviews, flags potential billing errors before payment, identifies upcoding patterns, and accelerates appeals processing. The scale of Medicare and Medicaid means even small efficiency gains generate hundreds of millions in savings or fraud prevention.

Military health benefits: The Defense Health Agency processes healthcare claims for active-duty service members, retirees, and dependents worldwide. AI handles claims routing, automates routine approvals, and identifies cases requiring human review based on complexity, cost, or clinical context.

Federal employee health benefits: OPM oversees health insurance for millions of federal employees and retirees. AI streamlines eligibility verification, automates premium calculations, and expedites enrollment processing during annual open season when volumes surge.

Claims Processing AI Workflow

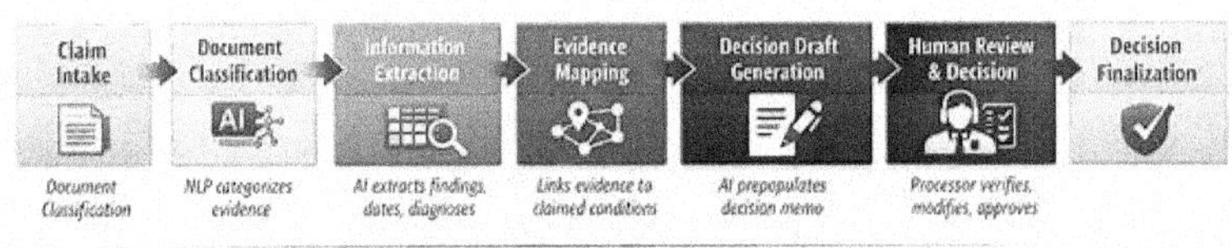

AI accelerates; humans decide.

Implementation Lessons From VA's Success

VA's claims automation success provides instructive lessons for federal healthcare managers considering similar deployments:

Start with high-volume, rules-based processes: Claims processing is ideal for AI because decisions follow documented criteria, evidence requirements are explicit, and the volume justifies automation investment. Avoid starting with low-volume, highly subjective processes.

Invest in data quality first: AI performance depends on structured, accessible data. VA spent years improving data quality, standardizing formats, and building extraction pipelines before deploying advanced AI. Clean data matters more than sophisticated algorithms.

Build trust through transparency: Processors adopted ADS because they understood what it did, how it worked, and why recommendations were made. Black-box systems that processors do not trust will not be used, regardless of technical performance.

Measure outcomes, not just efficiency: VA tracked not only processing time but also decision accuracy, appeal rates, and fairness across demographic groups. Faster processing that produces more errors or biased outcomes is not success.

Plan for change management: Processors worried AI would eliminate their jobs. VA leadership clearly communicated that AI was a productivity tool, not a workforce-reduction strategy. Workforce concerns must be addressed directly and honestly.

Iterate based on user feedback: VA continuously refined ADS based on processor input—adding features, fixing errors, and improving usability. Deployment is not a one-time event; it is an ongoing partnership with end users.

6.3   Scheduling Optimization: Maximizing Access and Minimizing Waste

Healthcare scheduling is deceptively complex. Clinicians have varying availability, different appointment types require different durations, patients have preferences and constraints, urgent cases need priority, and no-shows create last-minute gaps. Manual scheduling struggles to optimize across all these variables simultaneously. The result is wasted capacity, long wait times, frustrated patients, and overworked schedulers. AI scheduling optimization tackles this complexity through predictive analytics, intelligent matching, and proactive gap-filling.

The No-Show Problem

No-shows waste billions in healthcare capacity annually. When patients miss appointments without notice, clinical slots go unused, providers sit idle, and other patients who could have been seen remain waiting. The traditional response—overbooking—creates its own problems when more patients show up than capacity allows.

AI reduces no-shows through multiple mechanisms:

Predictive modeling: AI analyzes patient history, appointment type, time of day, day of the week, weather, distance to the facility, and dozens of other variables to predict the probability of a no-show for each scheduled appointment. High-risk appointments receive enhanced confirmation and reminder protocols.

Personalized reminders: Rather than sending identical reminders to all patients, AI determines optimal reminder timing and channel for each individual. Some patients respond best to text messages two days before appointments. Others prefer phone calls the day before. Personalization increases reminder effectiveness.

Intelligent overbooking: Instead of blanket overbooking policies, AI overbooks selectively based on predicted no-show rates for specific appointment slots. High-no-show-risk slots are overbooked; low-risk slots are not. This maximizes utilization without creating overflow chaos.

Proactive waitlist management: When cancellations occur, AI immediately identifies waitlisted patients who can fill the slot based on availability, appointment type match, geographic proximity, and clinical need. Automated outreach fills gaps within hours rather than leaving slots empty.

Barrier identification and intervention: AI flags patients with transportation challenges, childcare constraints, or other access barriers that increase the risk of no-shows. Care coordinators can proactively address barriers—arranging transportation, rescheduling to more convenient times, or converting to telehealth.

Healthcare organizations implementing AI scheduling optimization report 20 to 30 percent reductions in no-show rates. This translates directly to increased patient access without building more facilities or hiring more providers. The capacity was always there—AI helps you use it.

6.4 Appointment Type Matching and Provider Selection

Not every patient needs the same provider or appointment duration. A routine medication refill requires less time than a new patient evaluation. A nurse practitioner can handle a straightforward follow-up, while a complex diagnostic workup needs a specialist physician. AI optimizes provider-patient matching by analyzing clinical need, provider expertise, appointment duration requirements, and scheduling constraints.

Advanced scheduling AI considers:

Clinical acuity and complexity: Triaging patients to appropriate care levels—primary care, specialist, urgent care, emergency department—based on symptoms, history, and risk factors.

Provider expertise and availability: Matching patients with providers whose training and experience align with patient needs while respecting provider schedule preferences and workload distribution.

Appointment duration prediction: Estimating required appointment length based on reason for visit, patient complexity, and historical data. This prevents rushed appointments and reduces overtime.

Geographic optimization: Minimizing patient travel time by recommending the nearest appropriate provider with available capacity, especially important for rural health systems with multiple dispersed facilities.

Continuity of care: Prioritizing appointment slots with the patient's established provider when clinically appropriate, while avoiding excessive delays for continuity at the expense of timely access.

Group appointment opportunities: Identifying patients who would benefit from group visits for chronic disease management, health education, or shared medical appointments where efficiency and peer support combine.

6.5   Surge Capacity Management

Patient demand fluctuates daily, weekly, and seasonally. Flu season, allergy season, holiday periods, and public health emergencies all create demand surges. Traditional scheduling allocates capacity uniformly regardless of predicted demand, leading to either excess capacity during low-demand periods or access bottlenecks during surges.

Predictive scheduling AI forecasts demand patterns weeks or months in advance based on historical trends, seasonal factors, community health data, and real-time leading indicators. This enables proactive capacity adjustments:

Dynamic clinic schedules: Automatically expanding high-demand appointment types during predicted surge periods and reducing low-demand types.

Staff scheduling optimization: Aligning provider schedules with forecasted demand to avoid overstaffing during slow periods and understaffing during surges.

Facility load balancing: Directing patients to facilities with available capacity when their preferred location is fully booked, improving system-wide access.

Telehealth flex capacity: Offering telehealth appointments as surge relief for conditions that can be safely managed remotely, preserving in-person slots for cases requiring physical examination.

Preemptive waitlist outreach: Contacting waitlisted patients in advance of anticipated cancellations or capacity increases to ensure new slots fill immediately.

AI Scheduling Optimization Ecosystem

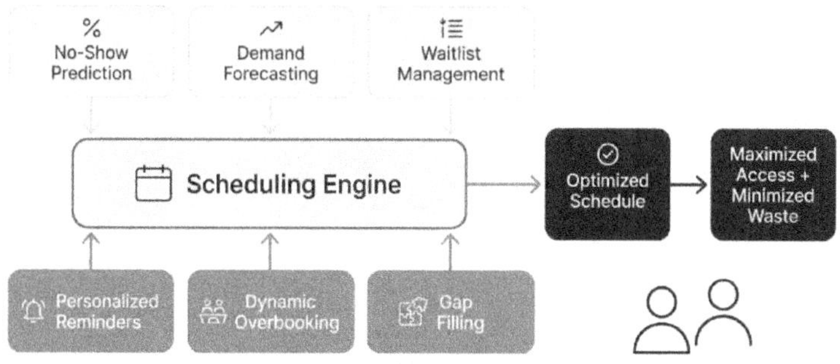

AI optimizes across multiple variables simultaneously.

Real-Time Schedule Adjustment

Even well-planned schedules face disruptions: emergency add-ons, providers running behind, equipment failures, and unexpected absences. AI-powered scheduling systems monitor real-time conditions and dynamically adjust:

Delay propagation mitigation: When a provider falls behind schedule, AI identifies which remaining appointments can be rescheduled with minimal patient impact and which patients to notify of delays.

Emergency insertion optimization: When urgent cases require immediate appointments, AI identifies the optimal insertion point that minimizes disruption to scheduled patients.

Provider substitution: When scheduled providers call in sick, AI recommends which appointments can be covered by available substitutes based on clinical requirements and which must be rescheduled.

Automated patient communication: Real-time notifications to affected patients about delays, rescheduling options, or alternative appointment slots without requiring scheduler intervention.

6.6   Contact Center Augmentation: Scaling Service Without Scaling Staff

Federal healthcare contact centers handle millions of interactions annually—appointment scheduling, prescription refills, benefits questions, insurance verification, test results inquiries, and general navigation support. Call volumes surge during open enrollment, public health emergencies, and policy changes. Traditional contact centers scale by hiring more agents, but recruiting, training, and retaining qualified staff is expensive and time-consuming. AI enables contact centers to handle dramatically higher volumes without a proportional increase in staff.

The Three Tiers of Contact Center AI

Effective contact center AI operates on three tiers with different levels of automation and human involvement:

Tier 1: Fully Automated Self-Service

AI chatbots and voice assistants handle routine inquiries without human involvement. Patients interact through website chat, mobile apps, or phone-based voice systems to accomplish straightforward tasks:

Appointment scheduling and rescheduling: Voice or chat interfaces guide patients through available slots, provider selection, and confirmation without agent assistance.

Prescription refill requests: Automated verification of patient identity, medication history, and refill eligibility with direct submission to pharmacy systems.

Insurance and eligibility verification: Instant confirmation of coverage status, copay amounts, and benefit details from integrated systems.

Test results access: Secure delivery of lab and imaging results with plain-language explanations and guidance on next steps.

FAQs and general information: Answering common questions about facility locations, hours, services, policies, and procedures.

Pre-visit information collection: Gathering patient history, symptoms, medication lists, and insurance information before appointments to reduce check-in time.

Tier 1 AI handles 40 to 60 percent of the contact center volume in mature implementations, freeing human agents to handle complex issues. The key to success is knowing when to escalate: AI must recognize when it cannot confidently resolve an inquiry and transfer seamlessly to human agents with full context.

Tier 2: Agent-Assisted AI

For inquiries requiring human judgment but benefiting from AI support, hybrid systems augment agent capabilities:

Real-time knowledge search: As agents converse with patients, AI surfaces relevant policies, procedures, and answers from knowledge bases without agents needing to search manually.

Next-best-action recommendations: AI suggests optimal responses or actions based on inquiry type, patient history, and successful resolution patterns from similar cases.

Sentiment analysis and escalation triggers: AI monitors the tone of conversations to identify frustrated or distressed patients who need supervisor intervention or specialized support.

Automated documentation: AI transcribes and summarizes calls in real time, reducing after-call work and allowing agents to focus on patient interaction.

Language translation: Real-time translation enables agents to serve patients in multiple languages without requiring multilingual staff at every shift.

Agent-assisted AI increases agent productivity by 25 to 40 percent while improving first-call resolution rates and customer satisfaction. Agents handle more calls in less time with better outcomes.

Tier 3: Human Specialist Escalation

Complex, sensitive, or unusual inquiries require human specialists. AI contributes by routing intelligently, providing specialists with comprehensive context, and documenting interactions:

Intelligent routing: AI analyzes inquiry content, patient history, and specialist expertise to route to the most appropriate available agent.

Context provision: Specialists receive AI-generated summaries of prior interactions, relevant patient history, and attempted resolution steps so they never ask patients to repeat information.

Quality assurance: AI reviews call recordings to identify training opportunities, policy compliance issues, and customer service excellence for coaching and recognition.

6.7   Conversational AI Design Principles

Building effective healthcare conversational AI requires careful attention to usability, privacy, and clinical appropriateness:

Natural language understanding: The system must handle medical terminology, abbreviations, typos, and varied phrasing. "I need to refill my blood pressure medicine," and "Can I get more of my BP meds?" should trigger identical workflows.

Context awareness: Conversations should maintain context across multiple turns. If a patient asks about test results and then asks, "When can I see the doctor about this?", the AI should understand that "this" refers to the test results.

HIPAA compliance: All patient interactions must meet federal privacy standards. Authentication is mandatory before disclosing protected health information. Conversations must be encrypted and audit-logged.

Graceful failure: When AI cannot understand or confidently respond, transfer to human agents smoothly with full conversation context. Never strand patients in automated loops.

Accessibility: Support for screen readers, voice input for patients with mobility limitations, and multilingual capabilities ensure equitable access.

Avoid clinical advice: AI should not provide medical advice, diagnostic suggestions, or treatment recommendations. Directing patients to appropriate clinical resources is safe; telling them what their symptoms mean is not.

Transparency: Patients should know they are interacting with AI and have easy access to human agents if preferred. Deceptive AI that pretends to be human undermines trust.

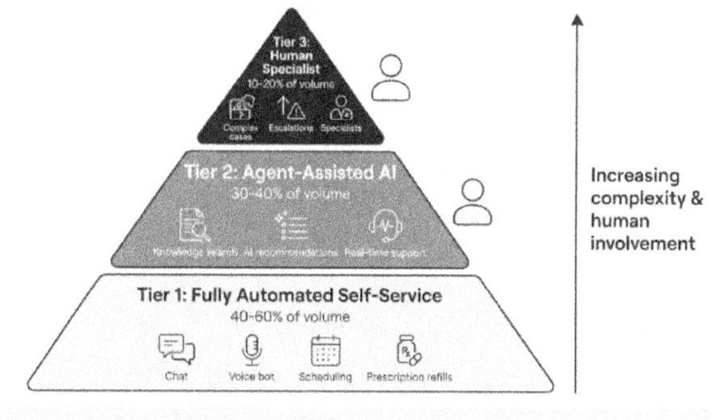

AI handles routine; humans handle complex.

Voice AI: The Next Frontier

While chatbots excel at text-based interactions, many patients—especially older adults—prefer phone calls. Voice AI brings conversational capabilities to traditional phone systems:

Natural conversation flow: Modern voice AI uses large language models to engage in natural dialogue rather than rigid menu trees. Patients describe their needs in their own words.

Interrupt handling: Unlike older interactive voice response systems, AI voice assistants handle interruptions naturally, allowing patients to clarify or change direction mid-conversation.

Emotion detection: Voice AI can detect frustration, confusion, or distress in tone and proactively offer escalation to human agents.

Multilingual support: Single systems supporting dozens of languages without separate implementations.

Integration with existing phone infrastructure: Deploying within the current call center technology rather than requiring complete replacement.

Early federal healthcare voice AI deployments focus on appointment scheduling, medication refills, and routine inquiries, with success rates exceeding 80 percent. As technology matures, capabilities will expand to more complex interactions.

Measuring Contact Center AI Success

Effective metrics for contact center AI include:

Containment rate: Percentage of inquiries fully resolved by AI without human agent transfer. Target: 40-60% for Tier 1 interactions.

First-call resolution: Percentage of inquiries resolved in a single interaction. AI should increase this by surfacing relevant information faster.

Average handle time: Time per inquiry including after-call work. Agent-assisted AI should reduce this by 20 to 35 percent.

Customer satisfaction: Patient ratings of interaction quality. AI should maintain or improve satisfaction relative to human-only service.

Agent productivity: Inquiries handled per agent per hour. Hybrid AI-human models typically improve this by 25 to 40 percent.

Escalation rate: Percentage of AI-initiated interactions requiring human takeover. Lower is better, but zero is unrealistic—AI should know its limits.

Abandonment rate: Percentage of patients who hang up before resolution. AI should reduce this by shortening wait times and improving self-service.

Supply Chain and Logistics: Ensuring the Right Resources at the Right Time

Healthcare supply chains are extraordinarily complex. Thousands of medical products with varying shelf lives, storage requirements, and usage patterns must be available precisely when needed. Stockouts delay surgeries and compromise patient care. Overstocking ties up capital and leads to waste through expiration. Manual inventory management struggles to balance these competing pressures across distributed facilities with fluctuating demand. AI optimizes this balance through demand forecasting, automated replenishment, and intelligent logistics.

6.8   Demand Forecasting and Inventory Optimization

Traditional inventory management uses simple reorder points: when stock drops below a threshold, order more. This approach ignores seasonal patterns, procedure volume trends, new treatment protocols, supplier lead times, and upcoming demand surges. AI demand forecasting analyzes historical usage data, scheduled procedures, clinical volume trends, seasonal patterns, and external factors to predict future needs with far greater accuracy.

AI-powered inventory systems provide:

Procedure-based forecasting: Predicting supply needs based on scheduled surgical cases, imaging studies, and clinical procedures rather than just historical averages. A week with many orthopedic surgeries requires different supplies than a week dominated by cardiac cases.

Seasonal and trend adjustment: Recognizing that flu season increases certain supply needs, summer trauma volumes differ from winter patterns, and gradual shifts in clinical practice affect long-term demand.

Lead time optimization: Accounting for supplier delivery schedules, transportation time, and order processing delays to ensure reorders arrive before stockouts despite variable lead times.

Multi-echelon optimization: Balancing inventory levels across central warehouses, facility storage, and department-level supplies to minimize total system inventory while maintaining availability.

Expiration management: Prioritizing use of items approaching expiration through first-expiry-first-out protocols automated through inventory tracking systems.

Automated reordering: Triggering purchase orders automatically when predicted demand and current inventory indicate future shortfalls, eliminating manual monitoring burden.

Healthcare systems implementing AI inventory optimization typically reduce inventory carrying costs by 15 to 25 percent while simultaneously decreasing stockout rates. The AI achieves what manual processes cannot: dynamic optimization across thousands of items simultaneously.

Supply Chain Resilience and Risk Management

The COVID-19 pandemic exposed healthcare supply chain vulnerabilities: single-source dependencies, just-in-time fragility, limited visibility into supplier tiers, and an inability to pivot rapidly when disruptions occur. AI enhances supply chain resilience through risk monitoring, supplier diversification analysis, and disruption response planning.

Risk management AI monitors:

Supplier financial health: Analyzing publicly available financial data, news, and market indicators to identify suppliers at risk of bankruptcy or operational failure before disruptions occur.

Geographic concentration: Flagging items where all suppliers or manufacturing sites are concentrated in single geographic regions vulnerable to natural disasters, geopolitical instability, or transportation disruptions.

Single-source dependencies: Identifying critical items available from only one supplier and recommending diversification or stockpile strategies.

Regulatory risk: Tracking FDA enforcement actions, quality issues, and recall patterns that may predict future supply disruptions.

Transportation vulnerabilities: Assessing how port closures, weather events, or logistics disruptions would affect supply availability and recommending mitigation strategies.

Alternative sourcing recommendations: When risks are identified, AI suggests alternative suppliers, substitute products, or process changes to maintain continuity.

Logistics Optimization

Once supplies are ordered, logistics determines how quickly and cost-effectively they reach points of use. AI optimizes transportation routing, warehouse operations, and last-mile delivery:

Route optimization: Calculating optimal delivery routes considering distance, traffic patterns, delivery time windows, vehicle capacity, and fuel costs. For federal healthcare systems serving remote or rural areas, route optimization can substantially reduce transportation costs.

Warehouse automation: AI-guided robots and automated storage and retrieval systems accelerate picking, packing, and inventory accuracy in central distribution centers.

Demand-driven distribution: Positioning inventory at facilities based on predicted need rather than uniform allocation, ensuring high-demand items are locally available while reducing overall system inventory.

Emergency supply deployment: During public health emergencies or natural disasters, AI rapidly recalculates optimal distribution to ensure critical supplies reach affected areas quickly.

Cross-facility sharing: Identifying opportunities to transfer excess inventory from facilities with surplus to those experiencing shortages, minimizing system-wide waste and stockouts.

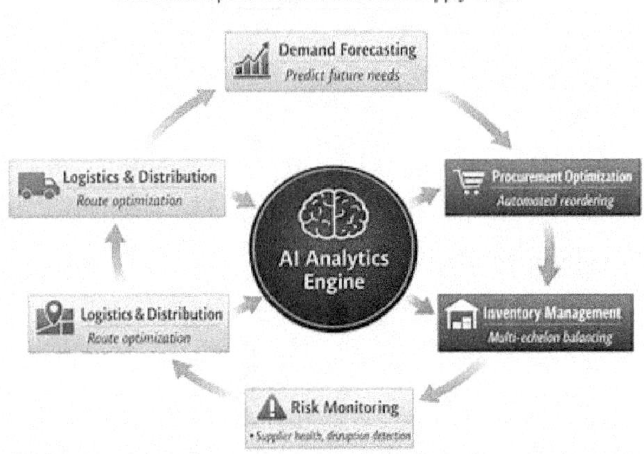

Clinical Supply Chain Integration

The most advanced supply chain AI integrates directly with clinical systems, creating closed-loop visibility from patient to product:

Procedure-based consumption tracking: Capturing exactly which supplies were used for each surgical case or clinical procedure, enabling precise cost accounting and evidence-based supply standardization.

Physician preference item management: Tracking which surgeons prefer which products and ensuring their preferred items are available while identifying opportunities for standardization to reduce inventory complexity.

Point-of-use inventory visibility: Real-time tracking of supplies in operating rooms, emergency departments, and clinical units so central supply knows exactly what is where without manual counts.

Automated charge capture: Ensuring supplies used for patient care are automatically documented and billed appropriately, eliminating revenue leakage from missed charges.

Evidence-based standardization: Analyzing clinical outcomes data to determine which products deliver the best results, supporting formulary decisions and reducing unnecessary variation.

Revenue Cycle Management: Capturing Every Dollar Owed

Revenue cycle management encompasses the entire process of capturing charges for services rendered, submitting claims to payers, resolving denials, collecting patient payments, and managing accounts receivable. Each step involves complex rules, tight deadlines, and costly errors. Incorrect coding reduces reimbursement. Denied claims delay payment. Missed charges lose revenue permanently. Manual revenue cycle processes are error-prone, slow, and expensive. AI transforms RCM through automated coding, denial prediction, intelligent claims scrubbing, and patient payment optimization.

6.9 Automated Medical Coding

Medical coding translates clinical documentation into standardized codes that determine reimbursement. Coders must understand complex clinical terminology, know thousands of diagnosis and procedure codes, follow detailed coding guidelines, and meet strict accuracy standards. Manual coding is slow, expensive, and error-prone. AI coding systems achieve over 95 percent accuracy while processing claims in minutes rather than hours.

AI medical coding capabilities include:

- Natural language processing of clinical notes: Extracting diagnoses, procedures, findings, and relevant clinical details from physician documentation, nursing notes, operative reports, and diagnostic studies.

- Automated code assignment: Selecting appropriate ICD-10, CPT, and HCPCS codes based on documented clinical information and coding guidelines.

- Real-time coding suggestions: Providing coders with AI-recommended codes they can accept, modify, or override, accelerating workflow while maintaining human oversight.

- Documentation improvement queries: Identifying missing or ambiguous documentation and automatically generating queries to physicians requesting clarification before claims submission.

- Compliance checking: Ensuring code combinations comply with payer-specific rules, medical necessity criteria, and regulatory requirements.

- Continuous learning: Improving accuracy over time by learning from coder edits, denial patterns, and updated coding guidelines.

Healthcare organizations implementing AI medical coding report 30 to 50 percent productivity improvements, coding backlog elimination, faster claim submission, and fewer denials due to coding errors. The financial impact is substantial: every day delay in coding and claim submission delays cash flow, and every coding error risks permanent revenue loss.

Denial Management and Prevention

Claim denials are expensive. Processing denials, submitting appeals, and managing resubmissions consumes staff time and delays revenue. Many denials are never successfully appealed, resulting in permanent revenue loss. AI transforms denial management from reactive appeals to proactive prevention.

Predictive denial prevention systems:

- Pre-submission claim scrubbing: AI reviews claims before submission to identify likely denial triggers—missing information, documentation gaps, medical necessity concerns, authorization issues, or coding errors. Fixing issues before submission prevents denials.

- Denial prediction modeling: Analyzing historical denial patterns to predict which claims are at high risk and why, enabling preemptive intervention.

- Authorization automation: Automatically determining which services require prior authorization, submitting authorization requests, and tracking approval status to prevent authorization-related denials.

- Medical necessity documentation: Identifying when clinical documentation must explicitly address medical necessity criteria to support reimbursement.

- Payer-specific rule application: Different payers have different rules, edit checks, and documentation requirements. AI applies the correct rules to each payer for each claim.

- Real-time eligibility verification: Confirming patient insurance coverage and benefit status at scheduling and registration to prevent denials due to eligibility issues.

- When denials do occur, AI accelerates resolution:

- Automated denial categorization: Classifying denials by reason code, payer, service type, and root cause to identify systematic issues requiring process improvements.

- Appeal prioritization: Focusing appeals staff on high-value denials most likely to succeed rather than wasting effort on low-probability or low-dollar appeals.

- Automated appeal generation: Drafting appeal letters with relevant clinical documentation, coding rationale, and policy citations for staff review and submission.

- Root cause analysis: Identifying why denials occur—billing errors, documentation gaps, authorization failures, coding mistakes—so processes can be improved to prevent recurrence.

6.10 Patient Payment Optimization

Patient financial responsibility has increased dramatically as high-deductible health plans become more common. Collecting from patients is harder than collecting from insurance because patients have more limited resources, less understanding of medical billing, and fewer legal obligations to pay quickly. AI improves patient collections while maintaining positive patient relationships:

- Payment propensity modeling: Predicting which patients are likely to pay quickly, which need payment plans, and which may require financial assistance, enabling tailored communication strategies.

- Optimal contact timing and channel: Determining when and how to contact patients about balances—phone, text, email, portal message—based on individual response history.

- Personalized payment plan offers: Automatically generating payment plan terms aligned with patient financial capacity rather than one-size-fits-all policies.

- Financial assistance screening: Identifying patients who likely qualify for charity care or financial assistance and proactively guiding them through application processes.

- Price transparency: Providing patients with estimated out-of-pocket costs before services when possible, improving payment rates, and increasing patient satisfaction.

- Self-service payment options: Mobile apps, text-to-pay, and online portals that make paying easy and convenient, reducing the need for staff phone calls.

AI accelerates and improves accuracy at every stage.

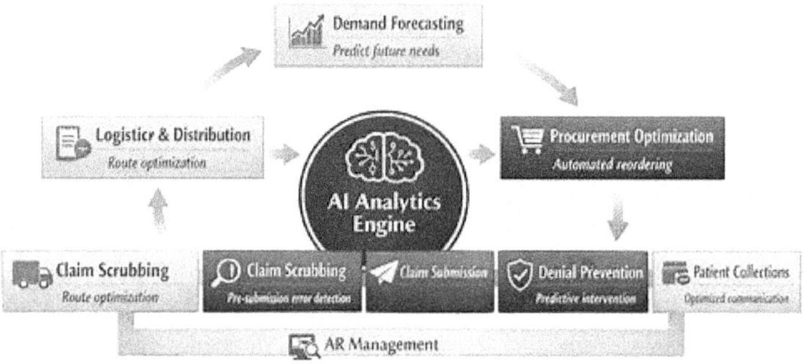

AI optimizes across multiple variables simultaneously.

Regulatory Compliance and Audit Support

Healthcare billing is heavily regulated. Federal payers enforce strict documentation, coding, and billing requirements. Violations result in audits, overpayment demands, penalties, and potential exclusion from federal programs. AI supports compliance through:

- Continuous coding audit: Automatically reviewing coded claims against documentation to identify potential compliance risks before external audits occur.
- Outlier detection: Flagging providers whose coding, billing, or utilization patterns differ significantly from peers, triggering internal review.
- Documentation sufficiency scoring: Assessing whether clinical documentation supports billed services and alerting providers to documentation gaps.

- Regulatory update monitoring: Tracking changes to coding guidelines, coverage policies, and billing rules and automatically updating compliance checks.

- Audit response acceleration: When external audits occur, AI rapidly retrieves relevant documentation, identifies similar cases, and supports response preparation.

6.11 Fraud, Waste, and Abuse Detection: Protecting Every Dollar

Healthcare fraud, waste, and abuse drain billions from federal programs annually. Fraudulent providers bill for services never rendered. Waste occurs through unnecessary services, inefficient processes, and duplicative testing. Abuse involves practices that violate reimbursement rules without criminal intent. Traditional fraud detection relies on manual sampling, retrospective review, and tips from whistleblowers—catching only a small fraction of improper payments. AI transforms fraud detection from retrospective recovery to proactive prevention through pattern recognition, anomaly detection, and network analysis.

The Scale of the Problem

In 2023, CMS reported 14.9 billion dollars in fraud, waste, and abuse recoveries. That number represents only detected and recovered improper payments—the true scale is certainly larger. Every dollar lost to fraud is a dollar unavailable for patient care. Every provider engaged in fraud undermines public trust in healthcare programs. Federal healthcare managers have both fiduciary and mission obligations to prevent, detect, and recover improper payments.

- Traditional Fraud Detection Limitations

- Legacy fraud detection approaches face fundamental limitations:

- Retrospective timing: Traditional audits occur months or years after improper payments, making recovery difficult and allowing fraud to continue undetected.

- Low sampling rates: Manual review capacity limits fraud investigators to tiny fractions of total claims, missing most fraud.

- Known scheme focus: Investigators look for fraud patterns they have seen before, missing novel schemes until tips or complaints surface them.

- Siloed data: Fraud spanning multiple programs, regions, or payers is hard to detect when each entity analyzes only its own data.

- Manual inefficiency: Human reviewers analyzing claims one by one cannot keep pace with fraud at scale.

- False positive burden: Broad audit criteria generate high false positive rates, wasting investigative resources on legitimate claims.

How AI Detects Fraud, Waste, and Abuse

AI fraud detection operates through multiple complementary techniques:

Pattern Recognition and Anomaly Detection

Machine learning models learn normal billing patterns for each specialty, procedure, geographic region, and patient population. Claims that deviate significantly from expected patterns trigger review. Anomalies include:

Impossible or improbable service combinations: Billing for procedures that cannot be performed together or services inconsistent with patient demographics.

Statistical outliers: Providers billing 10 times the volume of their peers or charging 3 times the typical rate.

Temporal anomalies: Claims patterns that suddenly change without clinical justification, such as rapid increases in high-reimbursement procedures.

Geographic inconsistencies: Urban providers billing extensive rural health visit codes or claims submitted from locations inconsistent with provider licensing.

Network Analysis

Fraud often involves networks of collaborating entities—providers, patients, pharmacies, medical equipment suppliers, and laboratories. Network analysis identifies suspicious relationships:

Referral rings: Providers who refer exclusively to co-conspirators regardless of patient need or geographic proximity.

Patient sharing patterns: Multiple unrelated providers billing for the same patients in coordinated patterns suggesting staged accidents or kickback schemes.

Corporate connections: Shell companies, shared addresses, interlinked ownership, or financial relationships among entities billing suspiciously.

Transportation schemes: Patterns where ambulance services, medical transportation, and specific providers consistently appear together suggesting kickback-driven unnecessary transport.

Behavioral Analysis

AI analyzes provider behavior over time to identify concerning trends:

Upcoding patterns: Systematic billing of higher-complexity codes than documentation supports.

Unbundling: Billing individual components separately to increase reimbursement when bundled codes apply.

Duplicate billing: Submitting the same claim multiple times to different payers or under different identifiers.

Service timing patterns: Billing for office visits during hours when the facility is known to be closed or the provider is documented elsewhere.

Patient volume impossibilities: Billing more patient encounters than physically possible, given time constraints.

Natural Language Processing of Clinical Documentation

When clinical documentation is available, NLP analyzes whether documented services support billed charges:

Documentation sufficiency: Whether clinical notes contain the details required to support procedure codes billed.

Copy-paste detection: Identifying documentation that is inappropriately copied across multiple patient records.

Medical necessity support: Whether documented clinical findings justify the services billed.

Template overuse: Detecting documentation patterns suggesting auto-filled templates rather than individualized patient care.

6.12 CMS Fraud Detection Success

CMS has deployed multiple AI fraud detection systems with measurable results. General Dynamics Information Technology developed the first production AI and machine learning model for CMS, identifying fraud schemes with more complex patterns than humans could detect. Over two years, the AI model outperformed manual methods in both detection accuracy and investigative efficiency.

In late 2025, CMS awarded Milliman for developing a glass-box AI model that proactively detects fraud, waste, and abuse in Medicare claims with unprecedented transparency. Unlike black-box models, the glass-box approach uses deterministic algorithms rooted in actuarial science that can be deconstructed to show exactly why providers were flagged. The system combines behavioral, network, and financial anomalies to generate single risk scores, allowing investigators to prioritize providers with both high billing costs and statistically abnormal patterns. This reduces false positives that drain investigative resources and focuses on providers presenting the greatest financial threat.

The measurable outcomes include faster identification of fraudulent providers, reduction in false-positive alerts that waste investigative capacity, and the ability to detect complex multi-entity fraud schemes that manual review misses.

AI finds needles; humans prosecute.

Explainable AI for Fraud Detection

Traditional AI fraud models faced adoption challenges because investigators could not understand why providers were flagged. Black-box models that output risk scores without explanation are difficult to use in legal proceedings and reduce investigator confidence. Explainable AI addresses this by showing:

Which specific claims or patterns triggered the alert?

How the flagged provider compares to peer benchmarks.

Which statistical anomalies are present and their magnitude?

Network connections to other suspicious entities.

Historical patterns that led to the current risk score.

Explainability serves multiple purposes: it helps investigators prioritize cases, supports legal proceedings with a documented rationale, builds investigators' trust in AI recommendations, and enables feedback loops in which investigators refine models based on case outcomes.

Proactive Prevention Versus Retrospective Recovery

The highest-value fraud detection prevents improper payments before they occur rather than recovering payments after the fact. Real-time or near-real-time fraud detection enables:

Pre-payment review: Flagging suspicious claims before payment for manual review, preventing improper payments rather than recovering them later.

Provider education: When patterns suggest billing errors rather than fraud, proactively educating providers prevents future improper billing.

Automatic edits: For well-established fraud patterns, automated claim edits can block payment entirely without manual review.

Continuous monitoring: Tracking provider behavior in real time allows intervention when concerning patterns first emerge rather than after extensive fraud has occurred.

Preventing one dollar of fraud is more valuable than recovering one dollar after payment because prevention avoids recovery costs, legal expenses, and the risk that fraudulent providers disappear before recovery is possible.

6.13 The Manager's Checklist: Deploying Administrative AI Successfully

Before deploying administrative and operational AI in federal healthcare, managers must ensure the following conditions are met:

Process Readiness

The current process is documented, measured, and understood before automation.

Pain points, bottlenecks, and inefficiencies are clearly identified.

Success metrics are defined with baseline measurements.

Process owners and end users support automation and are involved in design.

Workflows can accommodate AI outputs without major redesign.

Data Quality and Access

Data required for AI is available, structured, and accessible.

Data quality is sufficient—incomplete, inconsistent, or outdated data undermines AI effectiveness.

Data governance enables the use of AI and protects privacy appropriately.

Integration with source systems is technically feasible.

Historical data are sufficient in volume to train and validate models.

Technology and Integration

AI solution integrates with existing systems—EHR, billing, scheduling, inventory, CRM.

APIs and data interfaces are documented and supported.

IT infrastructure can handle computational and storage requirements.

Security and compliance requirements are met—FedRAMP, FISMA, HIPAA.

Vendor has proven federal healthcare experience and references.

Change Management and Training

Stakeholders understand why AI is being deployed and what it will do.

End users receive hands-on training before go-live.

Champions are identified to support peers and answer questions.

Feedback mechanisms allow users to report problems and suggest improvements.

Workforce concerns about job displacement are addressed honestly.

Performance expectations are realistic—AI augments, does not eliminate human work.

Governance and Oversight

Accountability is clear—who owns the AI system and its outputs.

Human oversight processes are defined where needed.

Audit mechanisms track AI decisions and outcomes.

Error reporting and escalation procedures exist.

Fairness monitoring ensures AI does not disadvantage specific groups.

Model performance is tracked continuously, and degradation triggers intervention.

Financial and Contractual

Total cost of ownership is understood to include licensing, implementation, training, maintenance, and monitoring.

ROI is calculated based on realistic assumptions.

Contracts include performance guarantees, service levels, and exit provisions.

Vendor lock-in risks are assessed and mitigated.

Budget includes not just AI software but also integration, change management, and ongoing operations.

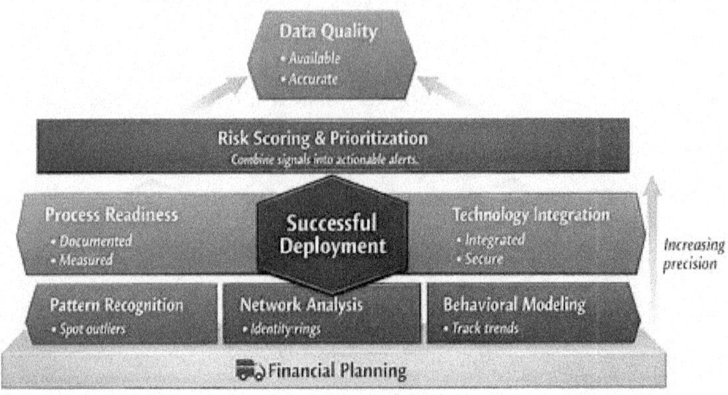

Chapter Summary

Administrative and operational AI may not carry the clinical stakes of patient-care applications, but its mission impact is undeniable. Every dollar of fraud prevented, every claim processed faster, every appointment scheduled efficiently, and every supply stockout avoided contributes directly to mission success. These are not peripheral applications—they are the operational backbone that enables healthcare delivery.

6.14 Key Takeaways

Claims processing automation, exemplified by VA's success in reducing processing time from 27 days to 12 hours, demonstrates AI's power to accelerate benefits delivery. The key is augmentation, not replacement—AI handles information extraction and organization while humans make decisions. This model achieved a 57 percent reduction in backlog and record productivity without sacrificing accuracy.

Scheduling optimization through AI reduces no-shows by 20 to 30 percent, maximizes clinical capacity utilization, and improves patient access without building new facilities or hiring more providers. The capacity was always there—AI helps organizations use it effectively through predictive analytics, intelligent matching, and proactive gap-filling.

Contact center AI operating on three tiers—fully automated self-service, agent-assisted support, and human specialist escalation—handles 40 to 60 percent of inquiries without human involvement while improving agent productivity by 25 to 40 percent for cases requiring human judgment. The key is knowing when to escalate and ensuring seamless transitions.

Supply chain AI optimizes across demand forecasting, inventory management, and logistics simultaneously—achieving what manual processes cannot. Organizations typically reduce inventory carrying costs by 15 to 25 percent while simultaneously decreasing stockout rates. Risk monitoring adds resilience by identifying supplier vulnerabilities before disruptions occur.

Revenue cycle management AI touches every stage, from automated coding, achieving 95% accuracy, to denial prevention through pre-submission claim scrubbing, to patient payment optimization. The cumulative financial impact is substantial: faster cash flow, reduced denials, improved collections, and better compliance.

Fraud, waste, and abuse detection through AI transforms from retrospective recovery to proactive prevention. CMS's explainable AI models identify complex fraud schemes that humans miss while reducing false positives that waste investigative capacity. Financial protection is measured in billions—CMS reported 14.9 billion in recoveries in 2023, representing only the detected cases.

Administrative AI succeeds when process readiness, data quality, technology integration, change management, governance, and financial planning all align. Deployment failures typically trace to gaps in one or more of these dimensions. The technology is necessary but not sufficient—organizational readiness determines outcomes.

6.15   What to Do on Monday Morning

If you are considering administrative AI, start with high-volume, rules-based processes where success is easily measured. Claims processing, scheduling, and contact center applications offer clear ROI and lower risk than attempts to automate complex, judgment-intensive work.

If you have administrative AI in operation, audit actual usage and outcomes. Are end users bypassing the system? Are promised efficiency gains materializing? Are there unintended consequences? Address issues based on data, not assumptions.

If AI performance is degrading, investigate causes. Data drift, workflow changes, policy updates, and system integration failures all degrade AI effectiveness over time. Continuous monitoring is not optional—it is an operational requirement.

If workforce resistance is blocking adoption, address concerns directly and honestly. Will jobs be eliminated? Will roles change? What support will be provided? Transparent communication builds trust; vague reassurances do not.

If you are building business cases, emphasize operational metrics that leadership understands, such as claims processing time, backlog size, no-show rates, inventory costs, denial rates, and fraud recoveries. Abstract AI capability matters less than concrete operational improvement.

If vendors promise unrealistic outcomes, demand proof. Case studies from similar organizations, references you can contact, pilot results from your own data, and performance guarantees in contracts separate credible vendors from those selling vaporware.

Administrative AI is not about replacing human workers—it is about eliminating the tedious, time-consuming work that prevents those workers from focusing on judgment, problem-solving, and mission delivery. When a VA claims processor handles nine times more claims, the veteran gets benefits nine times faster. When scheduling, AI fills appointment gaps, allowing patients to get care sooner. When fraud-detection AI identifies schemes that humans miss, taxpayer dollars go toward care rather than theft.

The managers who succeed with administrative AI are those who understand that technology deployment is the easy part. The hard part is process redesign, change management, stakeholder alignment, data preparation, and continuous improvement. The AI does not succeed on its own—it succeeds when organizations prepare for it, integrate it properly, train people to use it, monitor it continuously, and improve it relentlessly.

Federal healthcare produces massive volumes of administrative work. AI does not make that work disappear—it makes that work faster, more accurate, and less burdensome. The mission impact is the point. Technology is simply the means.

7 Population Health and Public Health AI

An epidemiologist at the CDC monitors syndromic surveillance data from emergency departments across the nation. It is early September, and the system is flagging an unusual pattern: influenza-like illness visits are spiking in the Southeast three weeks earlier than historical seasonal trends. The volume is not yet alarming—just 15 percent above baseline—but the geographic concentration and timing are atypical.

Forty-eight hours later, AI-powered disease surveillance confirms the signal across multiple data streams: pharmacy purchases of over-the-counter flu medications are elevated in the same regions, school absenteeism rates are climbing, and social media monitoring detects increased mentions of flu symptoms. The AI model estimates that, without intervention, cases will double every 5 days and spread to adjacent states within 2 weeks.

The epidemiologist does not wait for laboratory confirmation. She triggers the early warning protocol: notifying state health departments, activating targeted communications to healthcare providers in affected areas, expediting antiviral distribution to regional stockpiles, and recommending early vaccination campaigns in high-risk populations. Two weeks later, when lab data confirms an aggressive influenza strain, the response is already underway. Hospitalizations peak 30 percent lower than the AI model predicted without intervention. The early warning system bought time—and saved lives.

This is public health AI operating at scale: detecting signals in noise, forecasting disease trajectories, enabling proactive intervention, and measuring impact across entire populations. This chapter examines the AI applications that protect population health—surveillance and early warning systems, social determinants of health modeling, epidemiological forecasting, and public health emergency response.

For each domain, it explains what works, what federal healthcare managers must understand about unique challenges, and how to deploy responsibly.

Why This Matters

Population health and public health AI operate at a fundamentally different scale than clinical or administrative applications. Where clinical AI serves one patient at a time and administrative AI optimizes organizational processes, public health AI protects entire communities, states, and nations. The stakes are collective rather than individual, the data volumes are massive, and the interventions affect millions.

For managers, public health AI presents unique challenges. Privacy concerns intensify when surveillance systems track population behavior. Equity issues worsen when interventions benefit some communities while leaving others behind. Data quality problems that lead to minor errors in individual predictions can cause systematic failures at the population level. And public trust—already fragile after pandemic-era controversies—can collapse if AI systems are perceived as surveillance tools rather than public health protections.

But the mission value is extraordinary. Early disease detection saves lives through timely intervention. Social determinants modeling identifies vulnerable populations before health crises occur. Epidemiological forecasting enables resource prepositioning that prevents system collapse during surges. Emergency response optimization gets the right resources to the right places faster. Public health AI is not about individual optimization—it is about protecting populations at scale.

Surveillance and Early Warning Systems: Detecting Threats Before They Spread

Traditional disease surveillance relies on laboratory-confirmed diagnoses reported by healthcare providers to public health authorities. This process takes days to weeks, during which infectious diseases spread undetected. AI-powered surveillance systems monitor multiple real-time data streams—emergency department visits, pharmacy sales, social media, search queries, environmental sensors—to detect disease outbreaks days or weeks before traditional methods.

7.1   Syndromic Surveillance: The Foundation

Syndromic surveillance monitors pre-diagnostic indicators—symptoms, behaviors, and healthcare utilization patterns—rather than waiting for confirmed diagnoses. Emergency departments, urgent care centers, and pharmacies continuously transmit data to public health agencies, where AI algorithms analyze patterns for anomalies signaling outbreaks.

The CDC's National Syndromic Surveillance Program uses AI to analyze real-time patient symptom data from emergency departments nationwide. The system tracks over a dozen syndromes: influenza-like illness, gastrointestinal illness, respiratory distress, rash, neurological symptoms, and others. When patterns deviate from expected baselines, the system generates alerts for public health investigation.

What makes syndromic surveillance effective is speed and breadth. Traditional surveillance captures only patients sick enough to seek care and healthy enough to get tested. Syndromic surveillance captures everyone presenting to emergency departments, creating a broader and earlier view of disease activity. The AI contribution is pattern recognition at scale—identifying statistically significant deviations across thousands of facilities and dozens of symptom categories simultaneously.

7.2   Multi-Stream Data Fusion

The most powerful early warning systems combine multiple data streams to increase signal strength and reduce false positives. A single stream—like emergency department visits—might spike for dozens of reasons unrelated to infectious disease: weather events, sporting events, holiday weekends. But when multiple independent streams all signal the same pattern simultaneously, confidence increases that a true outbreak is emerging.

AI-driven multi-stream surveillance integrates:

Clinical data: Emergency department visits, urgent care encounters, primary care visits with relevant diagnoses or symptoms.

Pharmacy data: Over-the-counter medication purchases, prescription fills for antivirals or antibiotics, changes in demand patterns.

Laboratory data: Test orders, positive results, unusual testing volumes for specific pathogens.

Environmental data: Water quality monitoring, vector surveillance, air quality measurements, and climate conditions.

Digital signals: Search engine queries for symptoms, social media mentions of illness, health-related website traffic.

Sentinel surveillance: Targeted monitoring of high-priority populations or locations—schools, nursing homes, military bases, detention facilities.

Workforce data: Absenteeism rates, sick leave patterns, productivity changes in specific sectors or regions.

AI models trained on historical outbreak data learn which combinations of signals reliably predict disease emergence and which are noise. This fusion approach dramatically reduces false positive rates that plagued early syndromic systems, where every flu season triggered dozens of meaningless alerts.

CDC's Recent AI Infrastructure Investments

CDC has quietly built a modern AI infrastructure designed to reshape how public health data is collected, analyzed, and acted upon. These investments are now delivering operational value:

TowerScout: Legionella Prevention

Legionella bacteria grow in cooling towers and cause outbreaks of Legionnaires' disease—a severe pneumonia with 10 percent mortality if untreated. Identifying cooling towers during outbreaks traditionally required manual scanning of aerial imagery, taking roughly four hours per geographic area. CDC's TowerScout AI tool identifies cooling towers from satellite and aerial imagery in approximately five minutes. This 48-fold speed improvement enables rapid source identification during outbreaks and helps build preventive maintenance registries for high-risk towers.

BEACON: Global Disease Tracking

Boston University's BEACON system, developed in partnership with the CDC and funded with 52 million dollars in federal support, uses AI and human expertise to track emerging diseases globally. The system alerts health officials to potential threats so they can take action more quickly. BEACON's primary goal is to reduce the time between disease reporting and response—the critical window when early intervention prevents local outbreaks from becoming international emergencies.

Temporal and Spatial Pattern Recognition

Disease outbreaks have characteristic temporal patterns—how quickly cases increase, daily versus weekly cycles, seasonal trends—and spatial patterns—geographic clustering, transmission along transportation corridors, urban versus rural distribution. AI models trained on historical outbreaks learn these signature patterns and identify emerging outbreaks by matching current data to known templates.

Advanced systems go further, predicting where outbreaks will spread next based on population movement, transportation networks, demographic characteristics, and environmental conditions. This forecasting enables proactive resource positioning: prepositioning medical supplies, alerting healthcare facilities in predicted spread paths, and targeting public health messaging to at-risk populations before disease arrives.

The Alert Threshold Challenge

Early warning systems face a fundamental trade-off: sensitivity versus specificity. High sensitivity catches every outbreak early but generates many false positives. High specificity reduces false positives but risks missing outbreaks until they are already large. Public health agencies must calibrate systems based on disease severity, available response capacity, and tolerance for false alarms.

AI addresses this through tiered alerting:

Watch status: Subtle pattern deviations that warrant monitoring but not immediate action. Automated emails to relevant health departments.

Warning status: Statistical anomalies meeting moderate confidence thresholds. Triggers enhanced surveillance, laboratory testing, and response planning.

Alert status: High-confidence outbreak detection requiring immediate investigation and intervention. Activates emergency operations centers and resource mobilization.

The appropriate threshold depends on context. For diseases with high mortality and limited treatment options—like Ebola—even low-confidence signals trigger maximum response. For common seasonal illnesses with effective treatments—like influenza—higher thresholds reduce alert fatigue while still providing adequate warning.

7.3   Social Determinants of Health: Addressing Root Causes of Health Inequity

Clinical care accounts for only 10 to 20 percent of health outcomes. The remaining 80 to 90 percent reflects social determinants— economic stability, education, social environment, neighborhood conditions, and healthcare access. Traditional healthcare focuses almost exclusively on the clinical 10 to 20 percent. AI enables systematic integration of social determinants into population health management, identifying vulnerable populations before health crises occur and targeting interventions to address root causes.

What AI Reveals About Social Determinants

Social determinants data exists in fragmented, unstructured, and non-standardized formats: free-text clinical notes, community surveys, census records, environmental databases, transportation systems, housing registries, and social service records. Extracting meaningful insights from this mess requires natural language processing, geospatial analysis, and predictive modeling—capabilities AI provides that manual methods cannot match at scale.

Generative AI models effectively extract social determinants from clinical documentation. Natural language processing identifies mentions of housing instability, food insecurity, transportation barriers, social isolation, financial stress, and dozens of other social factors documented in clinician notes, social work assessments, and care coordination records. This converts previously invisible social context into structured, analyzable data.

Mass General Brigham's research demonstrated that generative AI models accurately highlight social determinants in doctors' notes, enabling systematic tracking of social risk factors across patient populations. This capability transforms how healthcare organizations understand the social context of health—from anecdotal awareness to quantified, actionable data.

7.4 Predictive Modeling for Targeted Interventions

AI predictive models incorporating social determinants achieve substantially higher accuracy than models using only clinical data. Studies show that incorporating social determinants into cardiovascular risk models improves the prediction of hospitalization, annual costs, and mortality. Models predicting readmission risk or chronic disease progression that ignore social factors miss critical drivers of health outcomes.

The practical value is targeting interventions to populations where they will have the greatest impact:

High-risk neighborhood identification: AI flags geographic areas where concentrations of adverse social determinants—poor air quality, food deserts, housing instability, limited healthcare access—predict elevated health risks. Public health resources can be targeted to these neighborhoods.

Individual risk stratification: Within clinical populations, AI identifies patients whose social circumstances place them at the highest risk for poor outcomes despite adequate clinical care. These patients benefit from care coordination, social work referrals, and connections to community resources.

Intervention matching: Different social barriers require different interventions. AI helps match patients to appropriate resources: transportation assistance for those with mobility barriers, nutrition programs for food insecurity, and housing support for patients experiencing housing instability.

Program evaluation: By tracking social determinants over time and linking them to health outcomes, AI helps evaluate whether social interventions actually improve health or merely consume resources without impact.

Health equity monitoring: AI measures whether healthcare quality, access, and outcomes differ systematically across populations with different social determinants, enabling targeted equity improvement efforts.

7.5   The Five Social Determinant Categories

The Healthy People 2030 framework organizes social determinants into five categories. AI applications address each domain:

Economic Stability

Employment status, income, financial security, and housing stability. AI identifies patients experiencing financial hardship by analyzing documentation, billing patterns, and public records. Interventions include financial counseling, enrollment in assistance programs, and payment plan optimization.

Education Access and Quality

Educational attainment, health literacy, and language barriers. AI detects health literacy challenges through communication patterns and tailors patient education materials to appropriate reading levels and languages. Predictive models identify populations requiring intensive education for chronic disease self-management.

Healthcare Access and Quality

Insurance coverage, access to primary care, transportation barriers, and healthcare utilization patterns. AI identifies populations underutilizing preventive services, experiencing access barriers, or receiving fragmented care. Geographic analysis pinpoints healthcare deserts requiring expanded services.

Neighborhood and Built Environment

Housing quality, environmental exposures, neighborhood safety, food access, transportation infrastructure. AI combines geospatial data with health outcomes to identify environmental health hazards. Models predict where asthma exacerbations, heat-related illness, or vector-borne diseases will emerge based on environmental conditions.

Social and Community Context

Social support networks, discrimination, social isolation, and community engagement. AI detects social isolation through healthcare utilization patterns, identifies populations experiencing discrimination through outcome disparities, and flags patients needing social connection interventions.

Converting fragmented social data into actionable insights.

Data Quality and Availability Challenges

Social determinants data faces unique challenges that clinical data does not:

Lack of standardization: No universal coding system exists for social determinants. Clinical data uses ICD-10 and CPT codes; social data is free text, local classifications, or absent entirely.

Documentation inconsistency: Clinicians document social factors haphazardly—some in structured fields, most in narrative notes, many not at all. Completeness varies wildly across providers and settings.

Privacy sensitivities: Patients may be reluctant to disclose housing instability, financial problems, or social isolation. Mandatory screening can feel intrusive. Documentation of sensitive social factors raises privacy concerns.

Cross-sector data access: Comprehensive social determinants require data from non-healthcare sectors—housing agencies, transportation systems, education departments, and social services. Data sharing agreements, technical integration, and governance alignment are complex.

Temporal dynamics: Social circumstances change frequently. A patient who was stably housed six months ago may be homeless today. Static snapshots miss these transitions.

Measurement challenges: How do you quantify social support or neighborhood safety? Many social determinants resist simple numeric measurement, yet models require quantifiable inputs.

These challenges mean that social determinants AI requires more data preparation, validation, and ongoing monitoring than clinical AI trained on standardized EHR data.

Ethical Considerations in Social Determinants AI

Using AI to identify vulnerable populations based on social characteristics raises ethical concerns that managers must address:

Stigmatization risk: Labeling patients as high social risk can lead to stereotyping, reduced expectations for health improvement, or discriminatory treatment.

Consent and transparency: Patients should understand that their social circumstances are being analyzed and used to guide interventions. Invisible algorithmic categorization undermines autonomy.

Intervention appropriateness: Identifying social needs is only valuable if interventions exist. Detecting food insecurity without connecting patients to food assistance creates awareness without benefit.

Avoiding determinism: Social risk scores predict population trends, not individual destinies. High-risk scores should trigger enhanced support, not fatalistic attitudes about inevitable poor outcomes.

Addressing root causes: AI that identifies social determinants without driving systemic change documents problems without solving them. The goal must be upstream intervention—changing social conditions, not just managing their health consequences.

Epidemiological Modeling: Forecasting Disease Trajectories

Epidemiological models predict how diseases spread through populations, enabling public health agencies to anticipate case volumes, project resource needs, evaluate intervention strategies, and communicate risk. Traditional models like SIR—Susceptible, Infectious, Recovered—use mathematical equations to describe transmission dynamics. AI-enhanced models incorporate far more complexity: heterogeneous populations, real-world contact networks, behavioral changes, intervention effects, and real-time data assimilation.

7.6   From Mathematical Models to AI-Enhanced Forecasting

Classic epidemiological models assume uniform populations: everyone has equal contact rates, identical susceptibility, and homogeneous behavior. These assumptions enable mathematical tractability but sacrifice realism. Real populations are heterogeneous: contact patterns vary by age, occupation, and location; susceptibility varies with prior exposure and comorbidities; behavior changes in response to perceived risk.

AI addresses this complexity through:

Agent-based modeling: Simulating individual agents with distinct characteristics, behaviors, and contact networks rather than treating populations as uniform masses. AI learns realistic agent behaviors from observational data.

Machine learning integration: Training models on historical outbreak data to learn transmission parameters, behavioral responses, and intervention effects rather than assuming them theoretically.

Real-time data assimilation: Continuously updating model parameters as new case data, testing data, and mobility data arrive, allowing forecasts to adapt as outbreaks evolve.

Ensemble forecasting: Running multiple models with different assumptions and combining their predictions to quantify uncertainty and improve accuracy—similar to weather forecasting.

Granular geographic modeling: Forecasting at the neighborhood or facility level rather than at the city or state level, enabling targeted interventions where they are most needed.

Scenario analysis: Rapidly testing thousands of "what if" scenarios—different intervention timing, intensity, or combinations—to identify optimal response strategies.

COVID-19: The Proving Ground

The COVID-19 pandemic became the largest real-world test of AI epidemiological modeling in history. Models guided lockdown decisions, hospital capacity planning, vaccination prioritization, and reopening strategies. The results were mixed—some models predicted accurately while others failed spectacularly—providing critical lessons for future modeling efforts.

What worked:

Short-term forecasting: Models predicting case volumes one to two weeks ahead achieved reasonable accuracy, enabling hospitals to prepare for surges.

Relative comparisons: Models comparing intervention scenarios— "lockdown reduces cases by 60 percent versus no intervention"— proved more reliable than absolute predictions.

Hospitalization forecasting: Predicting hospital bed and ICU needs one to three weeks ahead helped systems avoid collapse by expanding capacity and transferring patients.

Vaccination impact assessment: Models quantifying the effects of different vaccination strategies on transmission and mortality guided prioritization decisions.

What failed:

Long-term predictions: Models forecasting months ahead consistently overestimated or underestimated cases as behavior, variants, and immunity evolved unpredictably.

Behavioral uncertainty: Models struggled to predict how populations would respond to interventions—would compliance be 90 percent or 50 percent? Behavioral uncertainty dominated prediction error.

Variant emergence: New variants with different transmission rates and immune escape invalidated model assumptions trained on earlier strains.

Over-confident communication: Early models presented single forecasts without adequate uncertainty quantification, creating false confidence and subsequent credibility loss when predictions missed.

7.7   Practical Applications Beyond Pandemics

Epidemiological modeling applies beyond crisis response to routine public health planning:

Seasonal Influenza Forecasting

CDC runs annual influenza forecasting challenges in which modeling teams predict the timing, intensity, and geographic spread of flu season. The best models now predict peak timing within one week and peak intensity within one category. This enables proactive vaccination campaigns, antiviral stockpiling, and the preparation of healthcare systems.

Vaccination Program Optimization

Models identify optimal vaccination strategies balancing coverage, cost, and disease reduction. For diseases such as HPV and hepatitis B, models determine which age groups to target, whether catch-up vaccination is cost-effective, and how coverage thresholds affect population-level protection through herd immunity.

Vector-Borne Disease Prediction

Models that incorporate climate data, vector surveillance, and human case reports predict where and when mosquito-borne diseases such as West Nile virus or dengue will emerge. This enables targeted vector control—spraying, source reduction, public education—before human transmission occurs.

Healthcare Demand Forecasting

Beyond infectious diseases, models forecast healthcare utilization for chronic conditions, mental health crises, substance use disorders, and injury patterns. This supports capacity planning, staffing optimization, and resource allocation across public health systems.

Epidemiological Modeling Complexity Layers

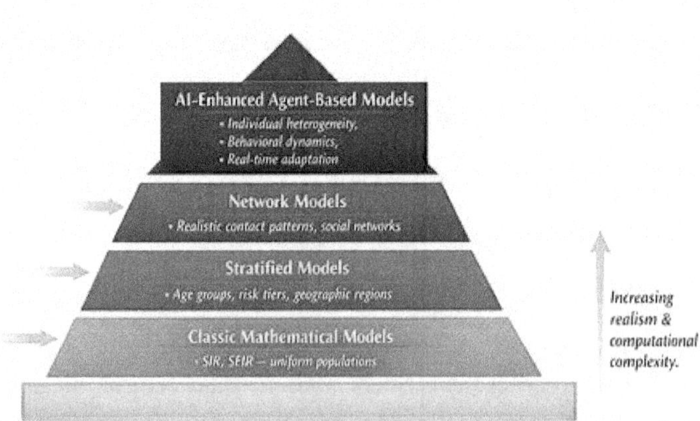

Higher layers capture reality better but require more data and compute.

Model Validation and Uncertainty Quantification

Epidemiological models are only valuable if decision-makers trust them enough to act on their forecasts. Trust requires rigorous validation and honest uncertainty communication:

Backtesting: Evaluating model performance on historical outbreaks where actual outcomes are known. Models that perform well on past data earn credibility for future forecasts.

Forecast scoring: Quantitative metrics like mean absolute error, calibration, and skill scores enable objective comparison of competing models.

Ensemble approaches: Combining multiple independent models and reporting the prediction range more accurately conveys uncertainty than single-model forecasts.

Scenario-based communication: Presenting "optimistic," "expected," and "pessimistic" scenarios with clear assumptions helps decision-makers understand the range of plausible futures.

Confidence intervals: Every prediction should include uncertainty bounds—not just a point estimate but a range reflecting model confidence.

Rapid updating: As new data arrives, models should update forecasts and transparently communicate when predictions change and why.

Assumption transparency: Decision-makers must understand what models assume about transmission rates, behavior, interventions, and immunity. Hidden assumptions undermine trust when violated.

Public Health Emergency Response: AI When Every Minute Counts

Public health emergencies—pandemics, natural disasters, environmental hazards, bioterrorism—compress decision timelines from weeks to hours. Traditional manual processes for situation assessment, resource allocation, communication, and coordination cannot keep pace. AI accelerates every stage of emergency response:

real-time monitoring, resource optimization, decision support, and coordination across fragmented systems.

7.8   Real-Time Situation Awareness

The first challenge in any emergency is understanding what is happening—scale, severity, geographic extent, and trajectory. Traditional situation reports aggregate data from dozens of sources through manual processes, producing summaries hours or days out of date. AI enables real-time situation awareness through automated data aggregation, analysis, and visualization.

WHO launched an AI-powered All-Hazard Information Management Toolkit in early 2026 that leverages generative AI to drastically reduce the time needed to produce critical situation reports during health emergencies. The toolkit automates data collection from multiple sources, generates synthesized assessments, and produces standardized reports that previously required extensive manual compilation. This frees emergency response staff to focus on decision-making and coordination rather than data compilation.

Real-time situation awareness systems integrate:

Case surveillance: Automated aggregation of case counts, demographics, severity, and outcomes from multiple reporting systems.

Healthcare capacity: Real-time tracking of hospital beds, ICU capacity, ventilator availability, and staffing across facilities and regions.

Supply inventories: Monitoring of medical supply stockpiles—personal protective equipment, medications, diagnostic tests—at local, state, and federal levels.

Transportation and logistics: Tracking supply shipments, personnel movements, and transportation network status.

Social media monitoring: Detecting early signals of emerging problems through social media analysis—supply shortages, healthcare access failures, misinformation spread.

Environmental monitoring: Real-time data from environmental sensors, weather systems, and hazard detection networks.

Geospatial visualization: Mapping case locations, resource distribution, vulnerable populations, and response activities for spatial situation awareness.

Resource Allocation Optimization

Emergencies create resource scarcity: insufficient hospital beds, limited supplies, overwhelmed healthcare workers, and inadequate testing capacity. Manual resource allocation relies on crude heuristics and cannot optimize across thousands of variables simultaneously. AI resource optimization achieves what manual processes cannot— optimal allocation considering supply constraints, demand forecasts, transportation logistics, equity considerations, and rapidly changing conditions.

During COVID-19, AI helped hospitals optimize the allocation of scarce resources such as ICU beds, ventilators, and personal protective equipment. Reinforcement learning algorithms developed adaptive allocation policies that adjusted in response to real-time data streams. Similar approaches optimized vaccine distribution networks to ensure critical doses reached priority populations efficiently.

AI resource optimization addresses:

Medical supply distribution: Determining optimal allocation of supplies across facilities based on current usage rates, forecasted demand, transportation constraints, and stockpile locations.

Healthcare worker deployment: Matching available clinical staff to facilities with the greatest need while respecting licensure requirements, expertise match, and workforce capacity limits.

Patient transfer coordination: Identifying when and where to transfer patients from overwhelmed facilities to those with available capacity, optimizing across geography, transport time, and clinical appropriateness.

Testing resource allocation: Prioritizing diagnostic testing when capacity is limited based on symptom severity, exposure risk, occupation, and public health surveillance needs.

Vaccine distribution: Optimizing cold chain logistics, allocation across demographics and geographies, and scheduling to maximize coverage while minimizing waste.

Mobile asset positioning: Determining optimal placement of mobile testing units, field hospitals, and emergency response teams to maximize geographic coverage and response time.

7.9   Decision Support Under Uncertainty

Emergency decision-makers face profound uncertainty: incomplete data, rapidly evolving situations, conflicting information, and high-stakes consequences. AI decision support systems synthesize available evidence, quantify uncertainty, project scenario outcomes, and present decision options with transparent trade-offs.

Effective emergency decision support provides:

Scenario forecasting: Projecting likely outcomes of different intervention options—lockdown versus testing-and-tracing, full reopening versus phased approach—with uncertainty bounds.

Trade-off analysis: Quantifying competing priorities—public health impact versus economic consequences, equity versus efficiency, short-term containment versus long-term sustainability.

Consequence prediction: Estimating downstream effects of decisions—how will closing schools affect workforce availability? How will supply chain disruptions cascade across healthcare systems?

Policy simulation: Testing proposed interventions in silico before real-world implementation to identify unintended consequences and optimization opportunities.

Evidence synthesis: Rapidly reviewing scientific literature, gray literature, and real-world data to inform evidence-based decision-making despite incomplete or conflicting information.

Decision documentation: Creating auditable records of what information was available, what options were considered, what was decided, and why—critical for after-action review and legal accountability.

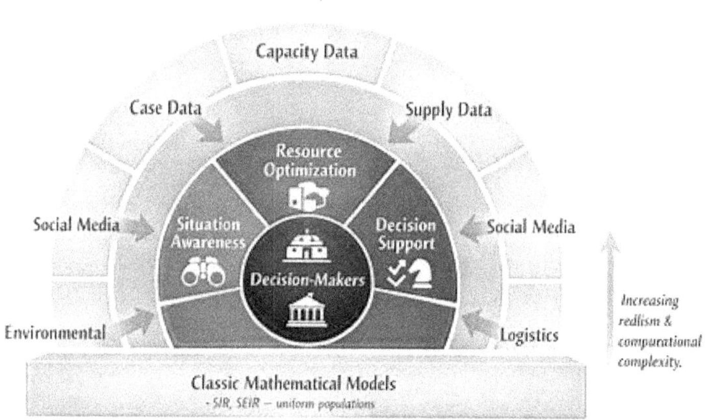

Higher layers capture reality better but require more data and compute.

Communication and Coordination

Public health emergencies involve dozens or hundreds of organizations operating simultaneously—federal agencies, state health departments, local authorities, healthcare systems, emergency management, law enforcement, and social services. Coordination failures are common: duplicate efforts, resource hoarding, conflicting guidance, and communication gaps. AI enables coordination at scale through automated information sharing, workflow orchestration, and collaborative platforms.

AI coordination capabilities include:

Automated status reporting: Healthcare facilities, supply distributors, and response teams automatically transmit status updates to centralized coordination platforms without manual data entry.

Request routing: When facilities request supplies, personnel, or assistance, AI routes requests to appropriate fulfillment sources based on availability, proximity, and capability.

Deduplication: Identifying when multiple organizations are addressing the same need, preventing wasted effort and enabling resource reallocation.

Gap identification: Detecting needs not being addressed by any organization, triggering assignment or escalation.

Communication synthesis: Generating situation summaries, status reports, and guidance documents from multiple data sources for distribution to stakeholders.

Conflict detection: Identifying when different agencies issue conflicting guidance or make incompatible resource commitments.

Workflow automation: Orchestrating multi-step processes—supply request, approval, shipment, delivery, confirmation—across organizational boundaries without manual handoffs.

7.10 Equity in Emergency Response

Public health emergencies often exacerbate existing health disparities. Vulnerable populations—low income, racial and ethnic minorities, limited English proficiency, disabilities, geographic isolation—face greater exposure risk, reduced access to interventions, and worse outcomes. AI can either mitigate or amplify these disparities depending on how it is designed and deployed.

AI risks that worsen equity:

Digital divide: AI systems that require smartphones, internet access, or digital literacy exclude populations without these resources.

Geographic bias: Resource optimization algorithms may systematically disadvantage rural or remote populations where allocation efficiency is lower.

Language barriers: Communication systems in English only exclude non-English speakers from critical information and services.

Algorithmic bias: Resource allocation or risk assessment models trained on historical data may perpetuate discriminatory patterns.

Invisible populations: Marginalized populations underrepresented in data sources may be invisible to AI systems, excluding them from interventions.

AI strategies that promote equity:

Equity-weighted optimization: Resource allocation algorithms explicitly prioritize vulnerable populations, accepting efficiency trade-offs to ensure equitable access.

Multi-modal access: Providing AI-enabled services through multiple channels—phone, web, mobile, in-person—to accommodate diverse capabilities.

Multilingual support: Natural language processing enabling communication in all languages spoken by significant populations.

Community-based data collection: Supplementing automated systems with community health workers and trusted community organizations that reach populations missed by digital systems.

Equity monitoring dashboards: Tracking whether interventions, resources, and outcomes are distributed equitably across populations, triggering corrective actions when disparities emerge.

Participatory design: Including representatives of vulnerable populations in AI system design to identify and address equity concerns before deployment.

Data Quality Challenges: The Foundation of Public Health AI

Public health AI relies on data from hundreds of sources, including hospitals, laboratories, pharmacies, schools, social services, environmental sensors, and public datasets. This data is collected by different organizations for different purposes, using different systems and standards. The result is fragmentation, inconsistency, incompleteness, and delays that undermine AI effectiveness. Data quality is not a technical detail—it is the foundation on which all public health AI stands or falls.

7.11   The Fragmentation Problem

During COVID-19, over one-third of local health agencies could not access surveillance data from local emergency departments. This failure was not technological—it was organizational. Healthcare systems, laboratories, and public health agencies operated on incompatible systems, with no integration, no data-sharing agreements, and no common standards. The patchwork nature of American public health created blind spots that allowed disease to spread undetected.

Fragmentation manifests as:

Technical incompatibility: Healthcare systems use dozens of different electronic health record platforms with proprietary data formats. Public health agencies cannot directly access this data.

Missing data-sharing agreements: Legal, privacy, and governance barriers prevent healthcare organizations from sharing data with public health agencies, even when technically feasible.

Inconsistent definitions: One jurisdiction defines "influenza-like illness" differently from another, making regional aggregation meaningless.

Incomplete coverage: Surveillance systems capture only facilities that participate voluntarily or are required by regulation. Many healthcare encounters—urgent care, retail clinics, telemedicine—are invisible.

Delayed reporting: Manual reporting processes create time lags of days to weeks between patient encounters and the availability of public health data.

Siloed programs: Different public health programs—immunization registries, disease surveillance, vital statistics—operate independent systems with no integration.

Data Quality Dimensions

Public health AI requires attention to multiple data quality dimensions:

Completeness

What percentage of relevant events are captured? Surveillance systems monitoring only large hospitals miss disease activity at small facilities. Immunization registries lacking data from retail pharmacies underestimate coverage. AI models trained on incomplete data produce biased predictions.

Timeliness

How quickly does data become available? Syndromic surveillance requiring real-time data loses value if reports arrive days late. Epidemic forecasting requires current data to project near-term trajectories. Delays render data historical artifacts rather than actionable intelligence.

Accuracy

How often are data values correct? Laboratory results with transcription errors, miscoded diagnoses, and incorrect demographic information all corrupt AI model inputs. Accuracy errors that average out at population scale may be tolerable; systematic errors that bias models are not.

Consistency

Do different sources define and measure the same concepts identically? If one facility codes shortness of breath as a respiratory symptom while another codes it as a cardiac symptom, syndromic surveillance breaks. Consistency enables aggregation across sources.

Granularity

How detailed is the data? Knowing influenza activity is elevated statewide helps minimally. Knowing which counties, which age groups, and which clinical severity levels are elevated enables targeted response. Granularity enables precision.

Strategies for Data Quality Improvement

Improving public health data quality requires coordinated action across technical, organizational, and policy dimensions:

Standards adoption: Implementing common data standards—FHIR for health data exchange, ICD-10 for diagnoses, LOINC for laboratory tests—enables interoperability.

Automated data exchange: Replacing manual reporting with automated electronic transmission from healthcare systems to public health agencies eliminates delays and transcription errors.

Data sharing agreements: Pre-negotiating master data sharing agreements covering emergency and routine public health uses so data can flow immediately when needed.

Data quality monitoring: Automated systems that flag missing data, impossible values, inconsistencies, and anomalies for investigation and correction.

Source diversification: Expanding surveillance beyond traditional healthcare to include pharmacies, schools, workplaces, and community organizations for broader coverage.

Data governance frameworks: Clear policies defining who can access what data for which purposes, with audit trails ensuring accountability.

Capacity building: Investing in public health data infrastructure—not just during emergencies but continuously—so systems are ready when crises occur.

Validation studies: Periodically comparing surveillance data against gold-standard sources to quantify accuracy and identify systematic errors.

7.12 Privacy and Surveillance Concerns

Public health surveillance using AI raises legitimate privacy concerns, especially following pandemic-era controversies over contact-tracing apps, location tracking, and vaccination passports. Public trust in public health—already fragile—collapses entirely if surveillance is perceived as government overreach rather than population protection.

Privacy-protective approaches to public health AI:

Aggregate reporting: Surveillance systems should report population-level patterns rather than individual-level data. Public health agencies need to know 50 people in a neighborhood have influenza, not the names of those 50 people.

De-identification: Removing personally identifiable information from surveillance data before analysis and storage.

Purpose limitation: Using surveillance data exclusively for public health purposes, never for law enforcement, immigration enforcement, or other non-health uses.

Transparency: Public communication about what data is collected, how it is used, who has access, and how privacy is protected.

Data minimization: Collecting only data necessary for public health purposes, not comprehensive individual profiles.

Access controls: Strict technical and policy controls limiting who can access surveillance data, with audit trails tracking all access.

Sunset provisions: Automatically deleting or anonymizing surveillance data after it is no longer needed for public health purposes.

Community engagement: Involving communities in decisions about surveillance systems, especially communities with historical reasons to distrust government data collection.

The goal is not maximum data collection—it is sufficient data for public health protection while respecting privacy, autonomy, and civil liberties. This balance is achievable but requires intentional design choices rather than defaulting to maximum surveillance.

7.13   The Manager's Checklist: Deploying Public Health AI Responsibly

Before deploying public health AI, managers must ensure the following conditions are met:

Surveillance and Early Warning Systems

Data sources are diverse enough to cross-validate signals and reduce false positives.

Real-time or near-real-time data feeds are established with participating facilities.

Alert thresholds are calibrated based on disease severity, response capacity, and tolerance for false positives.

Response protocols are defined for each alert level—who gets notified, what actions are triggered, what resources are mobilized.

Historical validation demonstrates that the system detects actual outbreaks earlier than traditional surveillance.

Public health staff trust the system enough to act on alerts rather than waiting for confirmation through traditional channels.

Social Determinants Integration

Social determinants data sources are identified and accessible.

Natural language processing or other extraction methods convert unstructured social data into analyzable formats.

Predictive models incorporating social factors are validated for accuracy and fairness across demographic groups.

Intervention resources exist to address identified social needs—identifying problems without solutions is futile.

Privacy protections prevent misuse of sensitive social information.

Patients are informed about social determinants screening and can decline without affecting care quality.

Equity monitoring tracks whether social determinants interventions reduce or perpetuate disparities.

Epidemiological Modeling

Models are validated on historical data where outcomes are known.

Uncertainty is quantified and communicated honestly—not single forecasts but ranges.

Assumptions are transparent and documented—decision-makers understand what models assume about transmission, behavior, and interventions.

Models update forecasts as new data arrives rather than producing static predictions.

Multiple models or ensemble approaches provide robustness against single-model failures.

Subject matter experts review model outputs before operational use.

Communication avoids over-confident predictions that will undermine trust when wrong.

Emergency Response Systems

Situation awareness automatically integrates data from multiple sources.

Resource allocation algorithms explicitly balance efficiency with equity.

Decision support provides scenario analysis with transparent trade-offs, not prescriptive answers.

Communication systems reach vulnerable populations through multiple channels and languages.

Coordination platforms connect all responding organizations with real-time information sharing.

Equity monitoring ensures vulnerable populations receive proportional resources and achieve equitable outcomes.

Privacy protections prevent misuse of emergency surveillance data for non-health purposes.

After-action review processes capture lessons learned for continuous improvement.

Data Quality and Governance

Data completeness is quantified—what percentage of relevant events are captured?

Timeliness meets operational requirements—data arrives when needed for decisions.

Accuracy is validated periodically against gold-standard sources.

Standards enable interoperability across data sources and jurisdictions.

Data sharing agreements are established before emergencies, not negotiated during crises.

Privacy protections are built into systems from design, not added as afterthoughts.

Public trust is monitored and maintained through transparency and community engagement.

Data infrastructure is funded sustainably, not just during emergencies.

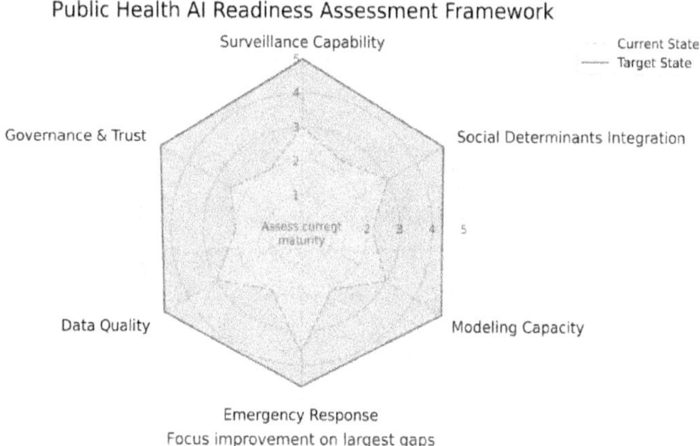

7.14 Chapter Summary

Population health and public health AI operate at a scale and impact that individual clinical or administrative applications cannot match. These systems protect entire communities through early disease detection, identify vulnerable populations before health crises emerge, forecast disease trajectories to enable proactive response, and optimize resource allocation during emergencies when every decision affects thousands. The mission value is extraordinary—but so are the technical, ethical, and governance challenges.

7.15 Key Takeaways

Surveillance and early warning systems using multi-stream data fusion detect disease outbreaks days to weeks before traditional laboratory-based surveillance. CDC's National Syndromic Surveillance Program, TowerScout AI, reducing tower identification time from four hours to five minutes, and BEACON global disease tracking demonstrate operational AI delivering public health protection at scale.

Social determinants of health account for 80 to 90 percent of health outcomes, yet healthcare has historically focused almost exclusively on clinical care. AI extracts social determinants from unstructured data, predicts which populations are at the highest risk, and enables targeted interventions that address root causes rather than manage health consequences. Models incorporating social factors substantially outperform clinical-only models.

Epidemiological modeling using AI achieves greater realism than classical mathematical models through agent-based simulation, machine learning integration, and real-time data assimilation. COVID-19 taught hard lessons about model limitations—long-term predictions failed, behavioral uncertainty dominated errors, and over-confident communication undermined trust. Short-term forecasting, scenario comparison, and honest uncertainty quantification remain valuable.

Public health emergency response AI accelerates situation awareness, optimizes resource allocation under scarcity, supports decision-making under uncertainty, and coordinates hundreds of organizations simultaneously. WHO's AI-powered emergency toolkit and COVID-era resource optimization demonstrated value—but also exposed data fragmentation and equity challenges that technology alone cannot solve.

Data quality challenges—fragmentation, inconsistency, incompleteness, and delays—are the primary barrier to the effectiveness of public health AI. During COVID-19, more than one-third of local health agencies could not access emergency department surveillance data in their own jurisdictions. Technical sophistication matters far less than data infrastructure, interoperability standards, and sustained investment in public health data systems.

Privacy and surveillance concerns require careful attention in public health AI. Aggregate reporting, de-identification, purpose limitation, transparency, and community engagement enable population protection while respecting civil liberties. The goal is sufficient surveillance for public health, not maximum data collection. Trust, once lost, is nearly impossible to rebuild.

Equity must be designed into public health AI from the start, not addressed after deployment. AI resource optimization can systematically advantage well-resourced populations at the expense of vulnerable communities unless equity is explicitly weighted. Digital divide, language barriers, geographic bias, and algorithmic bias all risk amplifying existing disparities unless proactively mitigated.

What to Do on Monday Morning

If you operate surveillance systems, audit data quality across all dimensions—completeness, timeliness, accuracy, consistency. Quantify gaps and prioritize fixes. AI built on bad data produces bad insights regardless of algorithmic sophistication.

If you are implementing social determinants screening, ensure that intervention resources are available before screening. Identifying housing instability without connecting patients to housing support creates awareness without benefit and may breed cynicism about whether healthcare organizations actually care.

If you use epidemiological models, demand transparency about assumptions, validation on historical data, and honest uncertainty quantification. Reject over-confident single-number forecasts that will undermine credibility when wrong. Insist on scenario ranges with explicit trade-offs.

If you are preparing for public health emergencies, establish data-sharing agreements, coordination platforms, and resource-allocation protocols now—not during crises. Emergency response speed depends on pre-existing infrastructure, not heroic improvisation when disasters strike.

If equity is not explicitly monitored in your public health AI systems, add equity dashboards immediately. Track whether surveillance coverage, resource allocation, and health outcomes are equitable across populations. Disparities invisible to dashboards persist uncorrected.

If public trust in your surveillance systems is uncertain, invest in transparency and community engagement. Publish what data you collect, how it is used, who has access, and how privacy is protected. Trust is built through consistent transparency, not marketing campaigns during crises.

If your jurisdiction experienced data infrastructure failures during COVID-19, advocate for sustained public health data funding. Infrastructure built during emergencies and defunded afterward will fail again. Continuous investment enables reliable systems when they matter most.

Public health AI is not about individual optimization—it is about protecting populations at scale. The technical challenges are substantial: integrating data from hundreds of fragmented sources, modeling complex disease dynamics, predicting human behavior, optimizing resource allocation under scarcity, and coordinating response across jurisdictional boundaries. But the non-technical challenges are greater: building data infrastructure that works across organizational silos, establishing trust in communities with historical reasons for suspicion, ensuring equity when algorithms naturally optimize for efficiency, and balancing public health protection with privacy and civil liberties.

The managers who succeed with public health AI are those who understand that technology is necessary but not sufficient. Data infrastructure matters more than algorithms. Interoperability standards enable capabilities that sophisticated models cannot. Trust built through transparency and community engagement determines whether systems are adopted or rejected. Equity designed into systems from the start produces different outcomes than equity added as an afterthought. And sustained investment in public health capacity—not just during emergencies—determines whether systems work when lives depend on them.

The next pandemic is not a hypothetical—it is inevitable. The question is not whether public health AI will be important—it already is. The question is whether we will learn from COVID-19's painful lessons and build public health AI systems that work when they matter most, that reach populations who need protection most, and that maintain the public trust on which all public health ultimately depends.

8   Fraud, Waste, and Abuse (FWA) Detection

A program integrity director at a federal health agency is reviewing the latest monthly report. On paper, everything looks fine: overall improper payment rates are flat, traditional audit programs are hitting their sampling quotas, and hotline tips are steady. But she knows those numbers don't tell the whole story. The fraud schemes that show up in audits are from last year. The most damaging fraud is almost always what you are not yet measuring.

On her screen is a new dashboard powered by an AI-driven FWA detection platform. One provider stands out—a home health agency in a mid-sized city that does not appear in any traditional fraud reports. Their overall billing volume is not extreme, but the pattern of services is strange. They are billing a very specific high-complexity visit code at a rate nine times that of similar agencies in the region. Almost all visits are documented as occurring on weekends. A network view shows tight connections to a small cluster of pharmacies and durable medical equipment suppliers with similarly unusual patterns.

The AI has scored this provider combination as high-risk, not because of any single data point, but because the collective pattern matches dozens of historical fraud cases across programs. The director flags the case for investigation. Within weeks, the investigation confirms a coordinated scheme: medically unnecessary services, kickbacks to patients, and staged equipment orders. Millions of dollars in improper payments are stopped. The fraud ring is dismantled. None of this would have surfaced through random sampling or human review of individual claims—the pattern was too complex, too subtle, and too distributed.

This is AI-powered FWA detection at work. It does not replace auditors and investigators. It gives them a radically better starting point. Instead of looking for needles in haystacks, they start where the hay is already smoking.

8.1 Why This Matters

Fraud, waste, and abuse are not abstract compliance issues—they are mission problems. Every dollar lost to fraud is a dollar that cannot be used for veteran care, military readiness, or services for vulnerable populations. Every unchecked abusive billing pattern erodes public trust in federal healthcare programs. Every missed opportunity to detect and prevent improper payments makes future oversight harder, not easier.

Traditional FWA programs rely on rules-based edits, retrospective audits, and whistleblower tips. These tools will always matter, but they do not scale to the volume, velocity, and complexity of modern claims and provider networks. AI changes the scale of what is possible. It can analyze billions of claims, spot non-obvious patterns across programs and geographies, and continuously learn from new schemes. But AI also introduces new risks—black-box models that investigators cannot defend in court, biased algorithms that over-target small practices while missing sophisticated corporate fraud, and alert overload that overwhelms limited investigative resources.

For serious AI managers in federal healthcare, FWA detection is a proving ground. It is one of the clearest, most measurable areas where AI can deliver workflow-level impact and mission-aligned value—if you design the operating model right. This chapter walks through how to do that.

8.2 Understanding Fraud, Waste, and Abuse in Federal Healthcare

Before you can deploy AI to detect FWA, you need a clear operational definition of what you are trying to find. In practice, frontline teams often lump everything under "fraud," but from a management perspective, the distinctions matter.

Operational Definitions

Fraud: Intentional deception or misrepresentation made with the knowledge that it could result in an unauthorized benefit. Examples include billing for services not provided, falsifying diagnoses to justify tests, or paying kickbacks for referrals.

Waste: Overutilization of services, or misuse of resources, not caused by criminal intent. Examples include redundant testing, inefficient care pathways, or failure to use generics when appropriate.

Abuse: Practices that are inconsistent with sound fiscal or clinical practices and that directly or indirectly result in unnecessary costs. Examples include upcoding visit levels, routinely ordering excessive diagnostics, or exceeding standard service frequencies.

AI is best at finding patterns associated with fraud and certain types of abuse. Waste is often a design and policy problem rather than a pattern-recognition problem. As a manager, you should be explicit about where AI will focus in your FWA portfolio and how you will handle the gray space between categories.

Where FWA Shows Up in the Workflow

Improper payments surface at multiple points across the healthcare and payment lifecycle. Mapping those touchpoints is a prerequisite for knowing where AI can help:

At the point of service: Medically unnecessary services, phantom visits, and misrepresented service locations.

In documentation: Copy-pasted notes, templated language that does not match billed complexity, and documentation that does not support billed services.

In coding and billing: Upcoding, unbundling procedures, duplicate billing, and misuse of modifiers.

In payment: Improper application of payment rules, incorrect coordination of benefits, or payments made despite known exclusions.

Across networks: Kickback arrangements, referral rings, and coordinated patterns among providers, pharmacies, labs, and equipment suppliers.

FWA Across the Payment Lifecycle

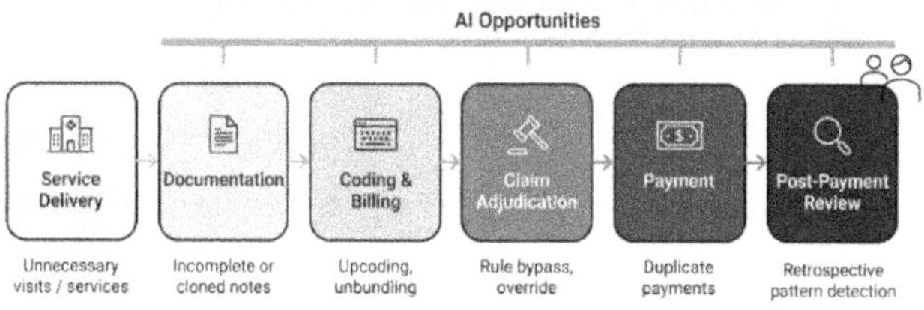

Understand where FWA can occur before deciding where AI should sit.

Core AI Techniques for FWA Detection

Under the hood, most FWA detection platforms use a relatively small set of core techniques. You do not need to be a data scientist to manage these systems effectively, but you do need to understand what these techniques do, what they cannot do, and how they show up in dashboards and alerts.

8.3 Rules and Expert Systems: The Baseline

Rules-based systems are where most agencies start and where many still spend most of their energy. These systems apply explicit conditions to claims—deny if age is under 18 and the procedure code is for a prostate exam, or flag if the same provider bills more than 24 hours of services in a day.

Strengths: Transparent, easy to explain, aligned with policy, defensible in audit or litigation.

Limitations: Easy for adversaries to adapt to, brittle when rules change, poor at catching novel or subtle schemes that fall between rules.

You should think of rules as the floor—not the ceiling. AI builds on these foundations rather than replacing them outright.

Supervised Learning on Known Bad Behavior

Supervised models learn from labeled examples—claims or providers that have historically been confirmed as fraudulent or abusive. The model learns patterns that distinguish good from bad and then scores new activities accordingly.

Use cases: Provider risk scoring, claim-level fraud likelihood, and pre-payment review prioritization.

Inputs: Claim attributes, coding patterns, amounts, timing, patient mix, prior audit outcomes.

Outputs: Risk scores for each claim or provider, often expressed on a 0–100 scale.

The main constraint is labels. Confirmed fraud cases are rare relative to all claims. That scarcity makes supervised models powerful in their training domain but fragile when applied outside it.

8.4　Unsupervised Anomaly Detection

Unsupervised models look for unusual patterns, not known bad ones. They ask which claims, providers, or networks look least like their peers without needing labeled fraud cases.

Use cases: Detecting new schemes, surfacing unusual behavior for analysts, scanning across entire universes of claims for outliers.

Techniques: Clustering, density estimation, autoencoders, isolation forests.

Outputs: Anomaly scores and groups of similar anomalies for investigation.

Anomaly detection is powerful but noisy. Not every outlier is fraud—some are legitimate edge cases or centers of excellence. Your operating model must assume that analysts, not algorithms, make final calls.

8.5   Network and Graph Analytics

Many of the most damaging schemes involve networks: a referring provider, a complicit lab, a friendly pharmacist, and a set of patients recruited for phony services. Graph analytics explicitly represents these relationships—nodes as entities, edges as relationships—and looks for suspicious patterns.

Use cases: Referral rings, collusive networks, shared ownership structures, repeated patient-provider-equipment triads.

Signals: Highly clustered subgraphs, unusually dense referral paths, and short path lengths among entities that should be weakly connected.

Network analytics is one of the clearest examples of AI doing something humans cannot do at scale. No investigator can hold a million-node network in their head. The graph engine can—if you feed it the right relationship data.

8.6   Hybrid and Glass-Box Models

Emerging best practice in federal healthcare is moving toward hybrid, glass-box models that combine statistical rigor with human interpretability. Rather than opaque deep learning models, these systems use structured scoring frameworks that show investigators exactly which factors contributed to a high-risk score.

For example, a provider risk score might be composed of:

Volume risk: How much is being billed relative to peers in the same specialty and region.

Coding behavior risk: Frequency of high-severity codes, rarely used modifiers, or odd code combinations.

Network risk: Connections to previously sanctioned entities, tight referral clusters, or shared addresses.

Behavioral risk: Sudden shifts in billing patterns, weekend-only services, or improbable service times.

Financial risk: Total dollars at stake and growth rate over time.

Each component is explainable on its own, and the combined score is a weighted sum of the components. Investigators can see not just that a provider is high risk, but why.

Linking Structured and Unstructured Data

FWA schemes rarely live entirely in structured fields. The most sophisticated actors know how to make their codes look plausible. The clues that matter are often buried in unstructured data, such as clinical notes, prior authorization narratives, appeal letters, call center transcripts, and even marketing materials. AI lets you mine these sources at scale.

8.7 Structured Data: Necessary but Not Sufficient

Claims tables, provider registries, fee schedules, and prior audit outcomes are the bread and butter of FWA analytics. They allow quick pattern scanning across billions of records. But they tell you what was billed, not what actually happened.

Claims fields: Procedure codes, diagnosis codes, modifiers, units, dates, places of service, provider identifiers, and amounts.

Provider data: Specialty, practice location, ownership, and sanction history.

Beneficiary data: Age, sex, geography, program enrollment type.

Any serious FWA AI program must master this structured landscape. But stopping here means you only see surface-level patterns.

Natural Language Processing for FWA

Natural language processing lets you analyze free text at scale. For FWA, the key use cases are:

Note-pattern analysis: Detecting templated or copy-pasted notes that do not change across patients or dates, suggesting documentation created to justify billing rather than record care.

Medical necessity validation: Comparing narrative documentation to billed codes to see if the story supports the service.

Appeal letter triage: Classifying appeal rationales and routing to appropriate teams, while spotting cases where the appeal story conflicts with clinical records.

Prior authorization narratives: Flagging patterns where justifications are identical across many patients, suggesting gaming of authorization rules.

Call-center transcripts: Identifying scripts used to recruit beneficiaries into questionable services or equipment orders.

Geospatial and Temporal Linking

Location and time data add critical context. For example:

Are providers billing for services at two distant locations simultaneously?

Are home visits clustered in a single building where many beneficiaries happen to require the same services?

Do service spikes align suspiciously with policy changes, payment increases, or natural disasters?

By combining structured claims, unstructured notes, and geospatial and temporal data, AI builds a richer picture of what is really happening than any single data source could provide.

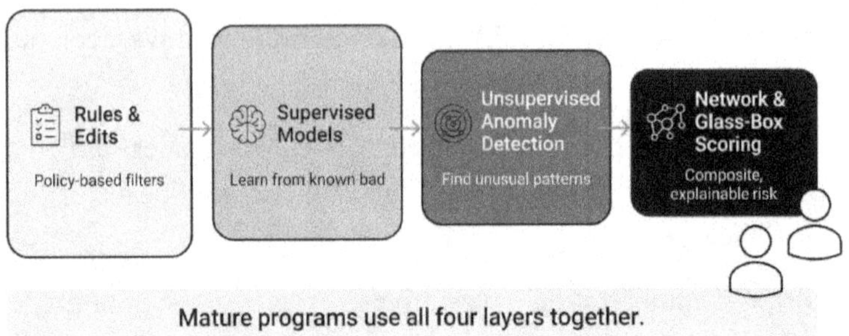

Guardrails for FWA AI: Avoiding New Kinds of Harm

FWA detection is inherently adversarial. As soon as you deploy new detection capabilities, sophisticated actors start probing for weaknesses. At the same time, most providers are honest. If your AI systems create a perception of a witch hunt—flagging large numbers of innocent providers with opaque scores—you will damage trust and invite backlash. Guardrails matter as much here as in clinical AI, just in different ways.

Explainability and Defensibility

Every FWA flag must be explainable to investigators, providers, and, when necessary, judges and oversight bodies. That means risk scores decompose into understandable components, each case includes a short narrative about the pattern detected and how it deviates from peers. Investigators can drill down from high-level scores to underlying claims and relationships.

8.8   Bias and Fairness in FWA

FWA models can encode and amplify bias. For example, if historical investigations disproportionately focused on small minority-owned practices, supervised models trained on that data will learn that providers like these are risky—even if the underlying risk is similar across groups. You must monitor and correct for this.

Monitor false-positive and false-negative rates across provider types and regions.

Check whether certain specialties or geographies are over-represented in high-risk flags after controlling for volume and case mix.

Audit sample cases from historically over-scrutinized groups to ensure the rationale is genuinely pattern-based.

Where feasible, include fairness constraints in model design to prevent demographic proxies from dominating scores.

8.9   Human-in-the-Loop for FWA

In FWA, human-in-the-loop means investigators, auditors, and program integrity staff always make final calls on escalations, payment holds, and enforcement actions. AI proposes; humans dispose.

No automated sanctions based solely on model output.

Tiered review levels based on risk bands.

Feedback capture from investigators into model retraining.

Clear documentation of human decisions for each high-impact action.

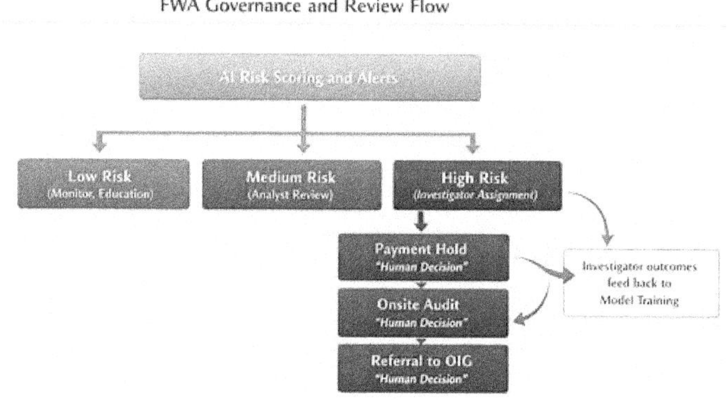

Case Management Integration: From Alerts to Actions

Many FWA AI pilots fail for a simple reason: they stop at generating alerts. Dashboards fill up with red flags that nobody has time or process to work on. The real value comes when AI is tightly integrated with case management workflows—assigning, tracking, and closing investigations in a disciplined way.

From Risk Scores to Work Queues

A usable FWA operating model translates analytics outputs into concrete work: provider-level queues sorted by risk and dollars, claim-level queues for pre-payment review, and thematic queues for patterns that demand policy or education responses rather than one-off investigations.

Workflow Integration

Effective integration with case management tools gives investigators a single view of all relevant data and AI rationales, tracks status and SLAs, standardizes closure reasons, and supports collaboration. Without this, AI becomes another dashboard that nobody truly owns.

8.10 Feedback Loop to Policy and Education

Not every anomaly is fraud. Sometimes patterns reflect confusing rules, poor communication, or unintended incentives. Your FWA operating model should route those findings to policy and education teams so systemic fixes accompany enforcement.

Without a closed loop, alerts become noise instead of action.

FWA Operating Model Maturity

Technology is only one dimension of an effective FWA program. Two agencies can buy the same AI platform and achieve completely different results. The difference is operating model maturity: how well data, analytics, governance, case management, and workforce practices fit together into a coherent system that actually changes outcomes.

8.11   Four Levels of Maturity

You can think about FWA AI maturity in four levels:

Level 1 – Rules-Only Reactive: Reliance on static edits, post-payment audits, and hotline tips. Little or no advanced analytics. Investigations begin long after payments are made.

Level 2 – Analytics-Enabled but Siloed: Some supervised or anomaly models exist, often built by a data science team, but outputs live in dashboards that are not embedded in case workflows. Investigators use them ad hoc.

Level 3 – Integrated Risk-Based Operations: AI risk scores drive formal work queues, thresholds are tuned to investigative capacity, case feedback feeds into model improvement, and governance guardrails are defined.

Level 4 – Adaptive, Cross-Program Intelligence: Multiple programs share signals where policy allows, network analytics reveal cross-program schemes, models adapt quickly to new patterns, and FWA insights routinely drive policy and education changes.

Most organizations today live between Levels 1 and 2. Moving to Level 3 and beyond is less about buying new tools and more about building the operating practices described throughout this chapter.

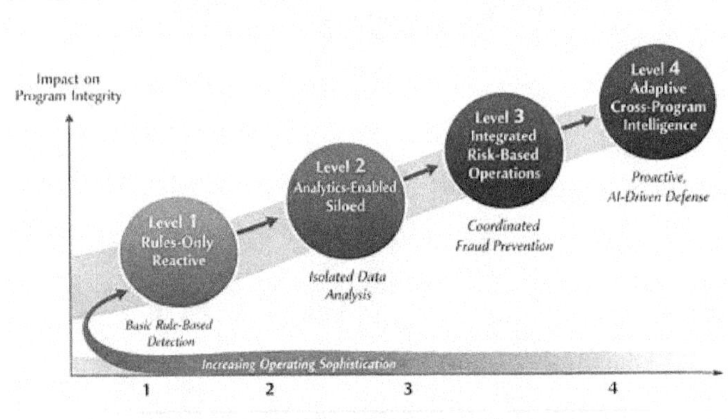

Without a closed loop, alerts become noise instead of action.

Lessons from OIG, CMS, and DoD

Federal oversight bodies have been experimenting with AI for FWA detection for several years. Their experiences offer concrete lessons for managers who want to avoid repeating the same mistakes.

8.12   What Has Worked

Targeted provider risk scoring to prioritize audits with higher hit rates per investigation.

Glass-box models that trade a small amount of statistical performance for much higher interpretability and legal defensibility.

Network analysis to uncover schemes that never appear in single-claim or single-provider views.

Cross-program analytics, where permitted, to detect schemes that hop between Medicare, Medicaid, TRICARE, and VA.

Embedding analysts with investigators so models evolve based on real-world investigative experience.

8.13   What Has Not Worked

Proof-of-concept models built in isolation with no clear operationalization path.

Alert deployments that flood investigators with more high-priority items than they can ever work.

Vendor black-box scores that cannot be explained or defended externally.

Assuming investigators will trust and use AI without training or early demonstration of value.

Building analytics only in reaction to scandals, then defunding when attention fades.

8.14   The Manager's Checklist: FWA AI That Actually Works

Use this checklist to assess whether your FWA AI program is manager-ready and responsible by design.

Strategy and Scope

Clear written objectives for what types of FWA the AI will target.

Prioritized focus areas based on value, feasibility, and risk.

Plain-language articulation of how FWA AI supports mission and program integrity.

Data and Infrastructure

Access to multi-year claims, provider, and beneficiary data with adequate quality.

A plan or capability to ingest and analyze unstructured data where it adds value.

Documented understanding of where cross-program linkages are possible and where they are prohibited.

Monitoring is in place for data drift and changing coding or policy patterns.

Models and Analytics

Use of a combination of rules, supervised learning, anomaly detection, and network analytics.

Explainable risk scores with component-level breakdowns.

Validation against historical cases, including fairness checks across provider types and geographies.

Regular refresh and recalibration based on new data and investigative feedback.

Governance and Guardrails

Documented policies for how AI outputs can and cannot be used.

Bias and fairness monitoring are embedded in regular governance reviews.

Clear human-in-the-loop workflows for significant enforcement decisions.

Ability to explain and defend AI-supported actions to internal and external oversight bodies.

Case Management and Operations

AI alerts feed directly into case management queues with clear owners and SLAs.

Investigators have unified views of relevant data and AI rationale inside their tools.

Investigation outcomes are coded in standardized ways and fed back into analytics.

Operational metrics tracked consistently: alert volumes, hit rates, time-to-first-touch, time-to-closure, and dollars prevented or recovered.

Workforce and Change Management

Investigators and auditors were involved in design and testing of AI tools.

Training covers how to interpret risk scores and when to override them.

Workforce concerns about AI replacing jobs are addressed with specific plans and examples of AI as a force multiplier.

Analytics champions exist within program integrity teams to bridge between data science and operations.

FWA AI Operating Model Blueprint

All columns must be built for FWA AI to deliver real-world impact.

Chapter Summary

FWA detection is one of the clearest places where AI can deliver concrete, measurable value in federal healthcare. It is also one of the easiest places to get AI wrong in ways that damage trust, overload investigators, and create legal and reputational risk. The difference is not the sophistication of your models—it is the maturity of your FWA operating model.

AI extends the reach of your program integrity teams. It spots non-obvious patterns across billions of claims, detects collusive networks that humans cannot see, links structured and unstructured data into richer pictures of behavior, and prioritizes limited investigative resources where they matter most. But AI does not decide who committed fraud or what enforcement actions to take. Humans still make those calls—and remain fully accountable for them.

The most successful FWA programs start with clear definitions and scoped objectives, combine multiple analytic techniques, invest in explainable models, integrate AI outputs tightly with case management, treat bias monitoring as a core requirement, and close the loop between investigations, policy, education, and model improvement. They move deliberately up the FWA operating model maturity curve rather than trying to jump from rules-only to fully autonomous analytics in one step.

For serious AI managers, FWA is both a test and an opportunity. Done well, it protects taxpayer dollars, strengthens program integrity, and supports honest providers, all while preserving due process and fairness. Done poorly, it produces dashboards nobody uses, overwhelms staff with noise, and invites scrutiny from the very oversight bodies you are trying to support. The tools are available. The difference will be how you design and lead the operating model around them.

Deep Dive: Building a Cross-Functional FWA Team

Program integrity lead: Owns overall strategy, prioritization, and accountability for results.

Data science lead: Owns model design, validation, monitoring, and technical roadmap.

Operations lead: Ensures case management workflows, SLAs, and staffing match alert volumes.

Legal and compliance: Reviews use of AI outputs for due process, evidentiary standards, and policy alignment.

Provider relations: Manages communication with providers, especially when AI-based reviews or audits are initiated.

Without this cross-functional structure, you will see "analytics in search of an owner" on one side and investigators complaining that models do not reflect reality on the other. The team composition matters at least as much as the specific platform you choose.

Deep Dive: Metrics That Actually Matter for FWA AI

Precision of high-risk alerts: Of all cases flagged as high risk, what percentage result in confirmed issues (fraud, abuse, material billing error)?

Coverage: What proportion of total spend and providers are meaningfully evaluated by AI each month, not just touched by rules?

Time-to-detection: How many months elapse between the start of a scheme and first meaningful flag?

Dollar impact: Dollars prevented or recovered attributable to AI-supported detection, net of program costs.

Behavior change: Reduction in targeted abusive patterns over time after AI-supported education or enforcement.

Executives will ask you whether the AI is "working." These metrics let you answer with specifics instead of anecdotes or model accuracy statistics that do not map to real-world value.

Deep Dive: Phased Implementation Roadmap

Phase 1 – Discovery: Inventory existing FWA tools, rules, and data sources. Stand up basic descriptive analytics and simple risk scores to demonstrate quick wins.

Phase 2 – Pilot: Select one or two well-scoped use cases (for example, a single specialty and region) and integrate AI outputs into investigator workflows with close monitoring.

Phase 3 – Scale: Expand coverage across specialties and geographies, tune thresholds to capacity, and formalize governance and retraining routines.

Phase 4 – Integrate: Connect FWA analytics with policy design, education planning, and procurement oversight so insights inform upstream decisions.

Phase 5 – Evolve: Introduce cross-program analytics, richer network models, and continuous learning as your team and infrastructure mature.

Trying to jump directly from Phase 1 to Phase 4 is how organizations end up with large contracts, impressive slideware, and very little change on the ground. Phasing your rollout lets you build trust, demonstrate value, and adjust course before you scale mistakes.

Deep Dive: Working with Vendors Without Losing Control

Own your data: Ensure contracts give you continued access to raw and derived data, not just dashboards.

Own your thresholds: You—not the vendor—should control how aggressive risk thresholds are, based on your capacity and risk tolerance.

Demand documentation: Require detailed model and feature documentation, including known limitations and performance by segment.

Plan for exit: Include provisions for model and configuration handover if you change vendors, to avoid starting from zero.

Insist on joint governance: Include vendor representation in governance forums—but keep decision rights inside the agency.

The goal is to treat vendors as capability accelerators, not as permanent black boxes. You need enough internal understanding to challenge, validate, and, if necessary, replace vendor models.

Deep Dive: Aligning FWA AI with Enterprise Risk Management

Map FWA risks in your enterprise risk register to specific AI controls and monitoring activities.

Use FWA analytics outputs as inputs into broader financial and operational risk dashboards.

Ensure board or executive risk committees see FWA AI metrics alongside traditional audit findings.

Document how FWA AI reduces specific risks—improper payments, regulatory findings, reputational damage—to justify ongoing investment.

Coordinate with internal audit so AI-supported findings complement, rather than duplicate, traditional audit work.

When FWA AI is positioned as a core control in your enterprise risk framework, it becomes easier to defend budgets, prioritize enhancements, and align multiple stakeholders around its importance.

9   Procurement Pathways for AI

A federal healthcare IT director is reviewing three proposals for AI-powered clinical decision support. Proposal A came through a traditional FAR Part 15 full-and-open competition—twelve months to award, an exhaustive technical evaluation, and custom terms negotiated line by line. Proposal B arrived via a GSA Multiple Award Schedule—pre-negotiated rates, three months to task order, but limited ability to customize beyond catalog offerings. Proposal C is pitched as an Other Transaction Authority prototype—fast, flexible, non-FAR—. Still, the legal team is nervous about intellectual property rights and whether it can transition to production without starting procurement from scratch.

All three vendors claim their AI models will improve diagnostic accuracy and reduce clinician burden. All three have plausible demonstrations. All three cost roughly the same over five years. But the procurement pathways are completely different, and the director knows that picking the wrong path will matter far more than picking the wrong vendor. The wrong pathway means years of delays, intellectual property disputes that block scaling, vendor lock-in that prevents competition, or compliance violations that attract OIG scrutiny.

The meeting ends with no decision. The director schedules another round of consultations with contracting, legal, IT security, and the CIO. Everyone agrees AI procurement is important. Nobody agrees on how to do it. Meanwhile, clinical teams continue using workarounds, and competitors in the private sector are deploying similar AI tools in months, not years.

This is the reality of AI procurement in federal healthcare. The technology is ready. The mission need is clear. The money is available. But the pathways are complex, the stakeholders are many, and the consequences of mistakes are severe. This chapter is your map through that complexity.

9.1 Why This Matters

Procurement is where AI strategy meets reality. You can have brilliant use cases, rock-solid governance, and enthusiastic stakeholders—but if you cannot acquire the AI system in a reasonable timeframe with reasonable terms, none of that matters. And in federal healthcare, procurement is not just complex. It is a minefield of regulations, policies, and stakeholder expectations that can turn a six-month project into a three-year ordeal.

The stakes are high. Choose the right procurement pathway and you accelerate mission delivery, maintain flexibility to adapt as AI evolves, protect intellectual property and data rights, preserve competition, and stay compliant with oversight requirements. Choose the wrong pathway and you face years of delays while competitors move ahead, vendor lock-in that prevents switching when better options emerge, intellectual property disputes that block you from improving or sharing AI models, compliance violations that attract investigations, and wasted taxpayer dollars on systems that never reach production.

For serious AI managers in federal healthcare, understanding procurement pathways is not optional. It is foundational. This chapter walks you through the major pathways, when to use each, what tradeoffs they involve, and how to navigate the process without losing your mind or your mission focus.

9.2 The Federal Procurement Landscape for AI

AI procurement does not happen in a vacuum. It sits at the intersection of multiple federal frameworks, each with its own rules, authorities, and oversight mechanisms. Before diving into specific pathways, you need to understand the broader landscape.

9.3 The Core Framework: FAR and Agency Supplements

The Federal Acquisition Regulation, or FAR, is the baseline. It governs most federal procurements and establishes the rules for competition, pricing, contract types, and contractor responsibilities. When people say a procurement is FAR-based, they mean it follows these foundational rules.

But FAR is just the start. Each major agency has supplements:

Defense Federal Acquisition Regulation Supplement (DFARS) for DoD, including TRICARE and Military Health System.

Veterans Affairs Acquisition Regulation (VAAR) for VA.

Health and Human Services Acquisition Regulation (HHSAR) for HHS agencies including CMS, NIH, CDC, and FDA.

These supplements add agency-specific requirements—cybersecurity standards, small business goals, compliance obligations, and approval authorities. You must comply with both FAR and your agency supplement.

9.4 The IT Overlay: FITARA and CIO Authority

The Federal Information Technology Acquisition Reform Act, or FITARA, changed IT procurement fundamentally. It elevated the CIO role, giving CIOs authority over IT budgets and acquisition decisions. For AI—which almost always involves IT systems, data, and infrastructure—that means your CIO must be involved early and often.

FITARA also requires agencies to conduct portfolio reviews, reduce duplication, and consolidate IT spending where feasible. That means you cannot just go buy an AI tool because your program office wants one. You have to demonstrate how it fits into the enterprise IT strategy, whether similar capabilities already exist, and why this acquisition makes sense at the portfolio level.

9.5   The Risk Overlay: OMB AI Guidance

Recent OMB memoranda—especially M-25-21 on AI use and M-25-22 on AI acquisition—impose additional requirements. M-25-22 specifically directs agencies to address intellectual property rights, privacy protections, vendor lock-in, performance monitoring, and ongoing testing in all AI contracts.

These requirements apply regardless of procurement pathway. Whether you use FAR Part 12, FAR Part 15, GSA Schedules, or Other Transaction Authority, you still have to include contract terms that protect government data rights, prevent vendor lock-in, and enable ongoing risk management.

9.6   Major Procurement Pathways for AI

There is no single best pathway for AI procurement. The right choice depends on your use case, timeline, risk tolerance, and organizational capacity. Here are the major options, with their strengths, limitations, and when to use each.

Federal AI Procurement Regulatory Stack

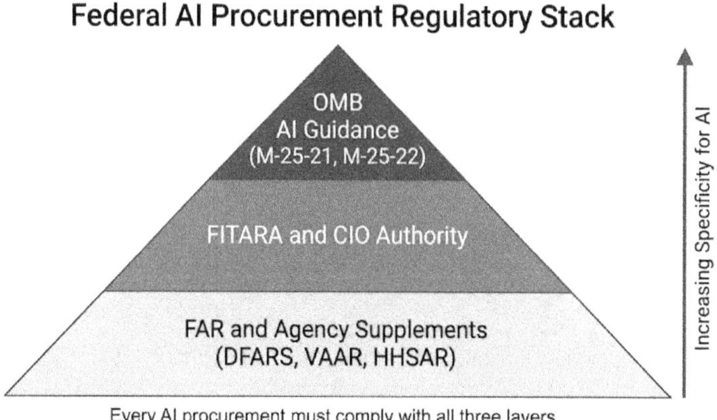

Every AI procurement must comply with all three layers.

FAR Part 12: Commercial Item Acquisition

FAR Part 12 treats AI as a commercial product or service—something already sold in the marketplace on commercial terms. This pathway prioritizes speed and simplicity by minimizing custom requirements and accepting vendor standard terms.

How it works: You conduct market research to confirm the AI system qualifies as commercial, issue a solicitation under simplified procedures, evaluate offers based on best value, and award a contract with streamlined FAR clauses tailored for commercial items.

Strengths:

Faster than full and open FAR Part 15 competitions—months instead of years.

Access to cutting-edge commercial AI that vendors would not sell under custom FAR Part 15 terms.

Lower administrative burden for both government and contractors.

Familiarity for vendors who sell AI commercially, reducing barriers to entry.

Limitations:

Limited ability to negotiate custom terms—vendors can refuse non-standard requirements.

Weaker data rights—government typically gets commercial licenses, not ownership or unlimited rights.

Less control over modifications, source code access, and future enhancements.

Risk of vendor lock-in if the commercial offering becomes the only viable option.

When to use: For AI systems that are widely available commercially, where customization is minimal, and where speed matters more than long-term control. Think enterprise AI platforms like Microsoft Azure AI, AWS AI services, or established SaaS AI tools.

FAR Part 15: Negotiated Acquisition

FAR Part 15 is the traditional full and open competition pathway. It allows extensive negotiation, custom requirements, and tailored contract terms. This is where agencies go when they need maximum control and are willing to invest the time to get it.

How it works: You develop detailed technical requirements, issue a request for proposals, conduct a thorough source selection with multiple evaluation factors, hold discussions and negotiations with offerors, and award a contract with fully customized terms.

Strengths:

Maximum flexibility to define custom requirements that fit your mission exactly.

Strong government rights to data, source code, and derivative works if negotiated properly.

Ability to structure payment terms, performance metrics, and compliance obligations exactly as needed.

Greater control over intellectual property, enabling government improvements and sharing across agencies.

Limitations:

Slowest pathway—often twelve to eighteen months from solicitation to award.

High administrative burden requiring substantial contracting and technical resources.

May deter some AI vendors who prefer faster, commercial-terms pathways.

Risk of over-specifying requirements before you fully understand what AI can do, leading to suboptimal solutions.

When to use: For mission-critical AI where you need custom integration with classified or sensitive data, long-term ownership and modification rights, or capabilities that do not exist commercially. Also use when the stakes justify the time investment—think AI for medical decision support systems that will be deployed across VA for decades.

GSA Multiple Award Schedules

GSA Schedules are pre-competed, pre-negotiated contracts that allow agencies to issue task orders quickly. Vendors get on the schedule through a one-time competition, then any agency can buy from them using simplified ordering procedures.

How it works: Verify that a vendor offering the AI you need is on a relevant GSA Schedule, review their catalog pricing and terms, issue a task order through GSA eBuy or directly with the vendor, and receive AI services under the pre-negotiated Schedule terms.

Strengths:

Very fast—task orders can be issued in weeks or a few months.

Pre-vetted vendors with government-friendly pricing and standard terms.

Reduced administrative burden compared to standalone procurements.

Competition still possible through multi-vendor task order competitions on Schedule.

Limitations:

Limited to what is already on Schedule—if no vendor offers your AI capability, this pathway does not work.

Less flexibility to negotiate terms beyond what Schedule allows.

Data rights and customization are constrained by Schedule contract terms.

Schedules can become outdated as AI technology evolves faster than Schedule refreshes.

When to use: For AI tools and services that are available from multiple Schedule vendors, where speed is important, and where standard commercial-like terms are acceptable. Common for AI professional services, cloud-hosted AI platforms, and standard AI software licenses.

Other Transaction Authority (OTA)

Other Transaction Authority allows certain agencies to enter agreements outside FAR entirely. OTAs are designed for prototype projects and research, offering maximum flexibility and speed but with significant constraints on transitioning to production.

How it works: An agency with OTA authority identifies a prototype or research need, works with an OTA consortium or issues a direct OTA solicitation, negotiates terms flexibly without FAR constraints, and awards an OT agreement for prototype development.

Who has it: DoD and DARPA are the largest OTA users. HHS agencies including NIH, CDC, and BARDA have OTA authority. DHS components including TSA and USCG can use OTAs. NASA and DOE also have OTA authority.

Strengths:

Fastest pathway—awards can happen in months, sometimes weeks.

Maximum flexibility in structuring agreements, IP rights, cost sharing, and performance terms.

Attracts nontraditional contractors, including startups and tech companies, who avoid FAR-based contracts.

Encourages innovation by reducing administrative overhead and allowing rapid iteration.

Limitations:

Prototype focus—OTAs are not for production systems. Transitioning prototypes to production often requires a separate FAR-based procurement.

Weaker oversight and accountability compared to FAR contracts, raising audit and compliance risks.

Intellectual property can be ambiguous—who owns improvements made during the OTA? What rights does government retain?

Not all agencies have OTA authority, and those that do face statutory limits on how it can be used.

When to use: For early-stage AI prototypes where the goal is proof of concept, not immediate production deployment. Use OTAs to test novel AI approaches quickly, then plan a FAR-based follow-on if the prototype succeeds. Do not use OTAs if you need production-ready systems immediately.

9.7   SBIR and STTR Programs

Small Business Innovation Research (SBIR) and Small Business Technology Transfer (STTR) programs fund R&D in small businesses, including AI innovation. These are not traditional procurements—they are grants or contracts specifically for technology development.

How it works: Small businesses respond to agency-specific SBIR or STTR topic solicitations, compete for Phase I feasibility awards (typically $250,000 over six to twelve months), successful Phase I awardees can compete for Phase II development awards (typically $1 to $1.7 million over two years), and exceptionally successful Phase II projects may transition to Phase III production contracts.

Who offers it: DoD, HHS (NIH, CDC, FDA), VA, DHS, DOE, NASA, and other major agencies all run SBIR and STTR programs with AI-related topics.

Strengths:

Access to innovative small businesses and startups developing cutting-edge AI.

Lower risk for government—small initial investment to test feasibility before larger commitments.

Supports economic policy goals by investing in U.S. small business innovation.

Potential pipeline to production if Phase I and II succeed.

Limitations:

Slow overall timeline—Phase I, II, and transition to production can take three to five years.

No guarantee of production transition—most SBIR projects remain R&D exercises.

Small business capacity limits—vendors may lack scale to deploy enterprise solutions.

Data rights can be complex, with vendors retaining significant IP developed under SBIR funding.

When to use: When you have a defined AI research problem but no existing commercial solutions, and you are willing to invest in R&D over multiple years with the understanding that production deployment is uncertain. SBIR works well for novel AI algorithms, domain-specific models, or high-risk technical approaches.

Procurement Pathway Comparison Matrix

	Speed			Control			Data Rights			Best For
	Fast	Medium	Slow	Low	Medium	High	Limited	Negotiable	Strong	
FAR Part 12										commercial off-the-shelf solutions
FAR Part 15										complex mission-unique systems
GSA Schedules										catalog-based buys
OTA										flexible prototyping
SBIR/STTR										small business innovation

No single pathway is always best—match pathway to use case and organizational capacity.

Critical Decision Points in AI Procurement

Choosing a pathway is only the first decision. Every AI procurement involves a series of make-or-break choices that determine whether you end up with a usable system or an expensive regret. These are the decisions where managers—not just contracting officers—must engage.

Data Rights and Intellectual Property

Data rights are the single most important and most misunderstood aspect of AI procurement. Get this wrong and you will discover—too late—that you cannot modify the AI, cannot share it with other agencies, cannot understand how it works, and cannot switch vendors without starting over.

The core questions:

Who owns the AI model and its training data?

What rights does government have to use, modify, and distribute the AI?

Can government share the AI with other agencies or state and local partners?

Does government have access to model weights, training data, and source code?

What happens to improvements made using government data during the contract?

In commercial AI procurements under FAR Part 12 or GSA Schedules, you typically get commercial licenses with limited rights. Vendors retain ownership. You get a license to use the AI as-is but cannot modify it, reverse engineer it, or redistribute it. That works fine for commodity AI tools but is problematic for mission-specific systems.

In negotiated procurements like FAR Part 15 or OTAs, you can negotiate for stronger rights—potentially including government purpose rights, unlimited rights, or custom arrangements. But vendors will resist, especially if they plan to commercialize the AI broadly. The negotiation becomes a trade between government control and vendor willingness to bid.

Managers rule: Decide early what rights you truly need. If the AI is purely a tool you will use without modification, commercial licenses may be fine. If you anticipate needing to adapt the AI, integrate it with other systems, or share it across agencies, insist on stronger data rights from the start. Trying to renegotiate data rights after award is nearly impossible.

Vendor Lock-In and Exit Strategy

AI systems create dependencies. Once deployed, they become embedded in workflows, training, and infrastructure. Switching vendors is costly and disruptive. That gives incumbents enormous leverage. Unless you plan ahead, you will find yourself locked in.

Common lock-in mechanisms:

Proprietary data formats that cannot be exported or converted.

AI models that require vendor-specific infrastructure to run.

Training data stored in vendor systems without export rights.

Integration architectures that tightly couple AI to vendor platforms.

Lack of documentation or knowledge transfer, making it impossible to operate the AI without the vendor.

OMB M-25-22 now requires agencies to include vendor lock-in protections in AI contracts. That means you must explicitly address data portability, model exportability, documentation standards, and transition assistance.

Practical protections:

Require vendors to export all government data in standard formats on demand.

Specify that AI models must be deployable on government-controlled infrastructure, not only vendor clouds.

Include contract terms requiring vendors to provide documentation sufficient for another party to operate the AI.

Structure contracts with shorter base periods and multiple option years, creating natural re-competition points.

Maintain internal expertise so you are not entirely dependent on vendor knowledge.

Managers rule: Assume you will eventually want to switch vendors. Design procurement and contract terms so that switching is possible, even if it is painful. Lock-in is a strategic risk, not just a procurement detail.

Performance Metrics and Acceptance Criteria

AI is probabilistic. It does not always work. It degrades over time as data distributions shift. Traditional acceptance criteria—does the system meet specifications—do not capture whether the AI actually performs well enough to be useful.

You need performance metrics that are measurable, mission-relevant, and testable:

Accuracy metrics: precision, recall, F1 score, or AUC-ROC for classification tasks.

Latency metrics: response time for predictions, especially in real-time clinical or operational use.

Fairness metrics: performance consistency across demographic groups, specialties, or facilities.

Robustness metrics: performance under edge cases, missing data, or unusual inputs.

Operational metrics: uptime, availability, and mean time to recovery from failures.

But metrics alone are not enough. You also need acceptance criteria that define what good enough means. For example, if your AI-powered diagnostic support must achieve 90 percent sensitivity and 85 percent specificity on a government-provided validation dataset before final acceptance, that must be in the contract.

Managers rule: Work with technical and clinical teams to define performance thresholds before you issue the solicitation. Make acceptance contingent on demonstrated performance, not just delivery of software. And build in ongoing performance monitoring requirements so you know if the AI degrades after deployment.

Security and Compliance from Day One

AI systems introduce new security and compliance risks. They ingest sensitive data, run on infrastructure that must be accredited, and make decisions that affect individuals. You cannot bolt security on after the fact.

Key requirements to address in procurement:

FedRAMP authorization or equivalent ATO processes for cloud-hosted AI.

HIPAA compliance for AI handling protected health information.

Authority to Operate timelines—specify who is responsible for obtaining ATO and when.

Audit and transparency requirements so you can understand how the AI makes decisions.

Incident response obligations if the AI is breached or malfunctions.

Testing and validation procedures, including adversarial testing for robustness.

One of the biggest procurement mistakes is assuming the vendor will handle security. In federal environments, ultimate accountability rests with the agency. If the AI does not get its ATO, the program stalls. If the AI violates HIPAA, the agency faces enforcement, not just the vendor.

Managers rule: Engage your IT security and privacy teams during requirements development, not after contract award. Specify security and compliance obligations explicitly in the statement of work and include penalties for non-compliance. Better yet, require vendors to demonstrate prior FedRAMP authorizations or ATOs for similar systems before you award.

AI Procurement Critical Decision Framework

Address all four decision points before issuing solicitation.

The Pilot-to-Production Problem

AI pilots succeed all the time. AI production systems are rare. The gap between proof-of-concept and operational deployment is where most federal AI efforts die. Procurement is a major reason why.

Why Pilots Fail to Scale

Pilots typically operate under special rules—small budgets, relaxed security, limited user bases, and temporary vendor arrangements. When you try to move to production, everything changes:

The pilot ran on a vendor sandbox with no ATO. Production requires FedRAMP or full ATO, adding six to twelve months.

The pilot used a small dataset that fit pilot-scale licensing. Production data volumes require enterprise licensing at much higher cost.

The pilot was funded through R&D or innovation budgets. Production requires sustained operations funding that competes with other priorities.

The pilot had a research-friendly contract vehicle like OTA or SBIR. Production requires a FAR-based vehicle that takes months to compete.

The pilot worked with enthusiastic early adopters. Production requires change management, training, and support infrastructure that nobody planned for.

The result: the pilot demonstrates value, leadership wants to scale, but the procurement and operational infrastructure to support production does not exist. The pilot stalls in bureaucratic limbo.

Designing Pilots for Production from the Start

The fix is not to skip pilots. Pilots are valuable for learning what works before committing to large-scale deployment. The fix is to design pilots with production in mind from day one.

Key principles:

Use production-like infrastructure for pilots—if production will require FedRAMP, pilot on FedRAMP-authorized platforms.

Include transition planning in pilot contracts—specify what happens if the pilot succeeds and how production will be procured.

Secure operations funding commitment before you start the pilot—leadership must agree that success triggers production investment.

Pilot with real users in real workflows, not just friendly testers in controlled environments.

Document everything—requirements, performance, lessons learned—so you can translate pilot insights directly into production solicitations.

Plan for a FAR-based follow-on even if the pilot uses OTA or SBIR—know which procurement pathway you will use for production.

Managers rule: Treat pilots as the first phase of production, not as separate experiments. Design contracts, infrastructure, and governance as if production is the default outcome for successful pilots. That mindset eliminates most pilot-to-production friction.

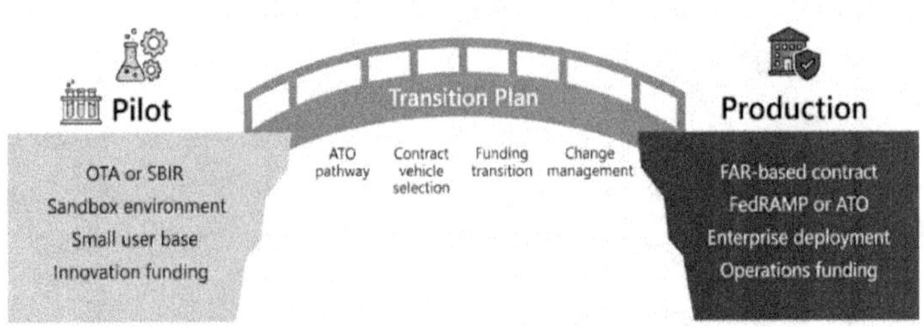

Build the bridge during pilot planning, not after pilot success.

Cross-Functional Coordination: Who Must Be at the Table

AI procurement is not a contracting problem. It is a cross-functional challenge that requires coordination across multiple stakeholders, each with legitimate concerns and authorities. Failure to coordinate is the second most common reason AI procurements fail, right after poor data rights planning.

Required Stakeholders

At minimum, every AI procurement requires these voices at the table:

Program or mission office: Defines the mission need, use case, and success criteria. Owns the funding and the operational deployment.

Contracting office: Ensures compliance with FAR, agency supplements, and OMB policies. Structures the solicitation and manages the source selection.

IT and cybersecurity: Ensures the AI integrates with existing infrastructure, meets security requirements, and can obtain necessary ATOs.

Chief Information Officer: Exercises FITARA authority, approves IT investments, and ensures alignment with enterprise IT strategy.

Privacy and civil liberties: Reviews AI for privacy risks, HIPAA compliance, and civil liberties impacts, especially for high-impact AI.

Legal and general counsel: Advises on data rights, intellectual property, liability, and compliance with all applicable laws.

Budget and finance: Confirms funding availability, structures payment terms, and ensures multi-year appropriations if needed.

In practice, these groups often operate in silos. The program office develops requirements without consulting IT. Contracting writes the solicitation without input from privacy. Legal reviews the contract only after award. By then, problems are locked in.

The Integrated Product Team Model

The best-run AI procurements use an integrated product team or IPT that brings all stakeholders together from the beginning. The IPT meets regularly, makes decisions collectively, and ensures that requirements, solicitation, evaluation, and contract terms reflect everyone needs.

IPT structure:

Program manager leads the IPT and owns the overall timeline and outcomes.

Contracting officer co-leads and ensures process compliance.

Technical lead from IT or data science defines technical requirements and evaluation criteria.

Security and privacy representatives review risks and approve security terms.

CIO representative ensures FITARA compliance and IT portfolio alignment.

Legal advisor provides real-time guidance on data rights, IP, and terms.

Budget representative confirms funding and structures payment approach.

IPTs meet weekly during pre-solicitation, biweekly during evaluation, and monthly post-award. The goal is shared understanding and joint ownership, not sequential handoffs where each group adds requirements that conflict with others.

Managers rule: Insist on an IPT structure for any AI procurement over $1 million or involving sensitive data or high-impact AI. Invest the coordination time up front. It is vastly cheaper than fixing problems after award or, worse, after deployment.

The Manager AI Procurement Readiness Checklist

Use this checklist before you issue any AI solicitation. Every no is a risk that will likely surface as a problem during or after procurement.

Mission and Requirements

You have a clear, written statement of the mission problem the AI will solve.

You have identified measurable success criteria and performance thresholds.

You have validated that AI is the right solution—not just the fashionable one.

You know what data the AI will use and have verified that data exists, is accessible, and meets quality standards.

Procurement Strategy

You have selected a procurement pathway based on mission need, timeline, and control requirements.

You have confirmed your agency has authority to use that pathway.

You have engaged contracting early and confirmed they have capacity to support your timeline.

You have identified whether this is a pilot or production procurement and planned accordingly.

Data Rights and IP

You know what data rights you need—use only, modify, share, or full ownership.

You have drafted contract language that secures those rights or confirmed that commercial licenses are acceptable.

You understand what IP the vendor brings and what IP will be developed under the contract.

You have addressed government rights to improvements made using government data.

Vendor Lock-In and Exit

You have identified lock-in risks—proprietary formats, vendor-specific infrastructure, lack of documentation.

Your solicitation includes requirements for data portability and model exportability.

You have specified transition assistance obligations if you switch vendors.

You have structured the contract with natural re-competition points via short base period and option years.

Performance and Acceptance

You have defined measurable AI performance metrics aligned with mission outcomes.

You have set acceptance thresholds that the vendor must meet before final payment.

You have included ongoing performance monitoring and reporting requirements.

You have addressed what happens if the AI degrades over time—retraining, updates, vendor obligations.

Security, Privacy, and Compliance

You have engaged IT security and confirmed ATO requirements and timelines.

Your solicitation specifies who is responsible for obtaining ATO and when.

You have addressed HIPAA, FedRAMP, and other compliance requirements explicitly.

You have included audit rights and transparency requirements so you can understand AI decision-making.

You have defined incident response and breach notification obligations.

Cross-Functional Coordination

You have established an integrated product team with program, contracting, IT, CIO, privacy, legal, and budget representation.

The IPT has met and agreed on requirements, evaluation approach, and timeline.

You have CIO approval or a clear path to approval under FITARA.

You have confirmed that funding is available and that multi-year appropriations are secured if needed.

Pilot-to-Production Planning if applicable

You have a written plan for how a successful pilot will transition to production.

You know which procurement vehicle you will use for production.

You have confirmed that production-scale funding will be available if the pilot succeeds.

The pilot is running on infrastructure that approximates production requirements.

You are documenting lessons learned in a format that can inform production requirements.

Assess readiness across all dimensions before issuing solicitation.

Practical Scenarios: Choosing the Right Pathway

To make procurement pathway selection concrete, here are five common federal healthcare AI scenarios with recommended pathways and rationales.

Scenario A: Enterprise AI Platform for Clinical Documentation

Need: AI-powered ambient clinical documentation that listens to patient encounters and generates draft notes, reducing clinician documentation burden across hundreds of facilities.

Recommended pathway: FAR Part 12 commercial acquisition via GSA Schedule if available, or direct FAR Part 12 if not.

Rationale: Multiple commercial vendors offer this capability. Speed matters because clinician burnout is urgent. You will use the AI largely as-is without deep customization. Commercial licensing with standard terms is acceptable because the AI does not expose unique government data and you can switch vendors if performance degrades.

Key contract terms: Performance SLAs for accuracy and latency. HIPAA compliance and BAA. FedRAMP or equivalent ATO. Data portability so notes and transcripts can be exported if you switch vendors. Clear vendor responsibilities for ongoing model updates as medical terminology evolves.

Scenario B: Custom Predictive Model for Suicide Risk Screening

Need: A custom AI model trained on VA patient data to predict suicide risk and prioritize outreach, integrated deeply with VistA and the new EHR.

Recommended pathway: FAR Part 15 negotiated acquisition with strong government data rights.

Rationale: This is mission-critical, sensitive, and requires custom development using government data. You need the ability to validate the model independently, modify it as clinical protocols evolve, and share it across VA without vendor dependency. Commercial off-the-shelf tools will not meet the need. The investment in a longer FAR Part 15 process is justified by the criticality and long-term nature of the deployment.

Key contract terms: Government purpose rights or unlimited rights to model, weights, and training code. Performance thresholds for sensitivity, specificity, and fairness across demographic groups. Ongoing model monitoring and retraining requirements. Explicit acceptance criteria tied to clinical validation. Source code escrow or delivery to protect against vendor failure.

Scenario C: Prototype AI for Fraud Detection in Pharmacy Claims

Need: A prototype AI system to detect unusual pharmacy billing patterns that may indicate fraud, waste, or abuse, tested initially in one region.

Recommended pathway: Other Transaction Authority prototype agreement, with a plan for FAR Part 12 or 15 follow-on if successful.

Rationale: The problem is well-defined but the best technical approach is uncertain. An OTA lets you test multiple algorithmic approaches quickly with a non-traditional contractor. The prototype timeframe is six to twelve months. If it works, you transition to a FAR-based production contract. If not, you learned cheaply what does not work.

Key contract terms: Clear definition of prototype success criteria. Government rights to use prototype results to inform future procurements. Explicit transition plan including data handoff and documentation if the prototype succeeds. No expectation of production deployment under the OTA itself.

Scenario D: AI Research for Novel Diagnostic Imaging Algorithm

Need: Exploratory research into AI algorithms that improve early detection of lung cancer from chest X-rays, with potential application across military and veteran populations.

Recommended pathway: SBIR Phase I and II.

Rationale: This is true R&D with high technical risk. SBIR provides a low-cost way to test feasibility with small businesses that may have novel approaches. Phase I confirms feasibility in six months for around $250,000. Phase II develops a working prototype over two years for up to $1.7 million. If successful, Phase III can transition to a production contract, potentially with the same vendor or through open competition.

Key contract terms: Deliverables focused on technical feasibility, algorithm performance on validation datasets, and documentation sufficient for independent evaluation. Negotiate government rights to use research results in future procurements. Accept that IP may be shared between government and contractor, typical of SBIR.

Scenario E: Scaled Deployment of Proven AI Workflow Tool

Need: You piloted an AI-powered appointment scheduling optimization tool successfully at five sites. Now you want to scale to two hundred sites across the enterprise.

Recommended pathway: If the pilot was OTA or SBIR, issue a new FAR Part 12 or FAR Part 15 solicitation incorporating lessons learned. If the pilot was already FAR-based, exercise option years or issue a follow-on task order.

Rationale: Scaling requires production-grade infrastructure, security, and support that pilots typically lack. You now know what performance looks like, what integration challenges exist, and what training and change management are required. Use that knowledge to write precise requirements. Compete the production procurement unless the pilot contract included well-structured options for scale.

Key contract terms: Performance metrics based on pilot results. Scaling milestones with payments tied to successful deployment at each tranche of sites. Training and change management support as contract deliverables, not afterthoughts. Exit strategy in case the vendor cannot scale effectively.

Procurement Pathway Selection Decision Tree

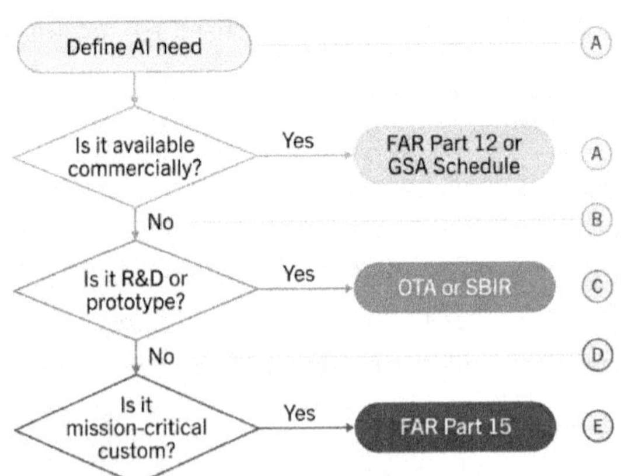

Match pathway to use case maturity and control needs.

Chapter Summary

Procurement is where AI ambition meets federal reality. You can have perfect use cases, flawless governance, and enthusiastic stakeholders, but if you cannot acquire AI systems at a reasonable speed, cost, and on reasonable terms, your AI program will stall. In federal healthcare, procurement is not just a process—it is a strategic capability that determines whether you lead or lag in AI adoption.

The federal procurement landscape for AI is complex. FAR and agency supplements provide the baseline rules. FITARA elevates CIO authority and demands portfolio-level thinking. OMB memoranda impose new requirements around data rights, vendor lock-in, privacy, and performance monitoring. Navigating this landscape requires understanding not just the rules but also the strategic choices those rules enable.

No single procurement pathway is always best. FAR Part 12 commercial acquisition offers speed and access to cutting-edge commercial AI but limits customization and data rights. FAR Part 15 negotiated acquisition provides maximum control and custom terms, but takes much longer. GSA Schedules offer a middle ground with pre-competed vendors and faster task orders. Other Transaction Authority enables rapid prototyping outside the FAR but is not designed for production. SBIR and STTR programs fund early-stage R&D, but transition to production is uncertain.

The critical decision points matter more than the pathway itself. Data rights and intellectual property determine whether you can modify, share, and evolve the AI over time. Vendor lock-in protections determine whether you have real competition or permanent dependency. Performance metrics and acceptance criteria determine whether you get AI that actually works or AI that technically meets specifications. Security and compliance requirements determine whether the AI can operate in federal environments at all.

The pilot-to-production gap is real and kills most federal AI efforts. Pilots succeed under special conditions that do not exist in production. The fix is to design pilots as the first phase of production, not as separate experiments. Use production-like infrastructure. Plan the production procurement vehicle before the pilot starts. Secure operations funding commitments up front. Document everything so pilot lessons translate directly into production requirements.

Cross-functional coordination is non-negotiable for AI procurement. Program offices, contracting, IT, CIO, privacy, legal, and budget all have legitimate authorities and concerns. Trying to sequence their involvement leads to conflict and rework. Integrated product teams that bring all stakeholders together from the beginning are the only reliable way to navigate complex AI procurements successfully.

For serious AI managers, procurement readiness is as important as technical readiness. Before you issue a solicitation, you must know what mission problem you are solving, what procurement pathway fits your need, what data rights you require, how you will avoid vendor lock-in, what performance you will accept, how you will achieve security compliance, who your cross-functional team is, and how pilots will transition to production if applicable. Miss any of these, and you will pay for it later.

The organizations that succeed with AI procurement are not necessarily the ones with the biggest budgets or the most sophisticated technical teams. They are the ones who understand that procurement is strategy, not paperwork. They match pathways to use cases deliberately. They design contracts that preserve flexibility and competition. They coordinate stakeholders early. They plan for production from day one. And they treat every procurement as an opportunity to learn and improve the process for the next one.

AI technology will continue evolving faster than procurement rules. New models, new vendors, and new capabilities will emerge constantly. What will not change is the need for disciplined, strategic thinking about how you acquire AI systems that work, that you can control, that you can evolve, and that you can defend to oversight bodies. Master that, and procurement becomes an enabler of AI adoption

10 Conclusion

Value-based healthcare transforms medical care while prioritizing patient-centered outcomes, cost-effectiveness, and equal healthcare access. Enhanced care delivery across all levels can be achieved by integrating financial data with social, clinical, and wellness data from healthcare systems. Medical organizations are dedicated to ensuring financial incentives match health quality improvement through innovative payment methods backed by patient savings accounts that leverage modern digital automatic payment systems. The strategic implementation of social support networks, including transportation access services, faith organizations, and governmental support services, transforms healthcare from illness treatment into a complete wellness enhancement approach. Through stakeholder collaboration, healthcare systems will develop sustainable, effective, and equitable solutions to redefine the future of health through value-based healthcare strategies.

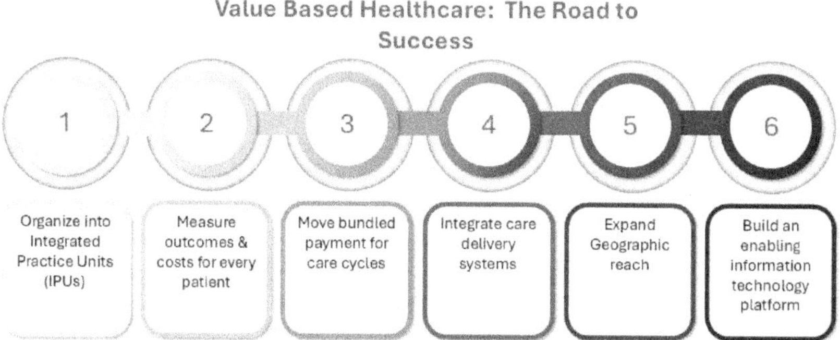

Figure 3 Road to Success

The healthcare sector has been under significant challenges in recent years. First, there are general challenges in healthcare delivery, such as ensuring patient satisfaction while fostering patient safety, financial challenges and rising healthcare costs, and advancements in healthcare technology, with associated challenges such as cybersecurity, regulatory standards, and the evolution of payment and invoicing systems. At the industry level, common challenges in healthcare include a growing human resource shortage, soaring patient expectations, pressure to deliver affordable care, and competition for skilled staff from foreign countries. These issues affect all healthcare organizations and may require more systemic interventions to reduce them. In rural areas, healthcare organizations face core challenges as well as other challenges due to the unique characteristics of those areas. Some of the unique challenges in rural healthcare include issues of access, affordability, health awareness, and quality. At the organizational level, rural healthcare faces challenges such as difficulties in attracting and retaining skilled manpower, the sparse distribution and limited reach of healthcare facilities, and low health awareness among local communities. These rural healthcare challenges are attributed to factors such as underfunding, poor social, human, and physical capital, and transportation and network constraints, among others. The disparities in healthcare challenges across different levels of service delivery point to the need for divergent interventions for the sector.

Illness Prevention

Figure 4 Illness Prevention

Attaining equitable care is a significant challenge for the U.S. healthcare system, and Medicare, through its emphasis on value-based care, seeks to address it to some extent. The burden of healthcare costs disproportionately affects different categories of citizens, particularly young adults, people with low income, adults with chronic conditions, those from minority populations, and the uninsured in the U.S. The U.S. healthcare system faces various challenges, including high healthcare costs (coupled with issues of non-insurance and under-insurance), low scientific and health education awareness among the general U.S. population, serious disparities in healthcare quality and outcomes, and chronic disease management, among others. The issue of healthcare costs is a particularly serious challenge as it remains a driver of even greater disparities in care quality and patient outcomes. Notably, Medicare has designed its insurance services to address these issues, particularly by focusing on goals such as ensuring access to basic medical care for underserved populations, including older people and people with disabilities.

As part of its initiative, the CMS, through Medicare, remains focused on promoting healthcare equity through various practices, including advocacy for value-based care models. Notably, VBCMs have the potential to improve healthcare equity by focusing on communication and coordination of care, as well as on patient treatment rather than on the disease. Some of the areas impacted by value-based care, and which have significant positive implications for equitable care delivery, include control of healthcare costs through streamlined registration for Medicare services, emphasis on team-based care and communication, treatment of the patient as a whole, focus on small patient groups with shared needs, supporting the role of physicians as healers without geographical limitations, and chronic disease management. The incentives provided as part of some VBCMs also provide intrinsic motivation for healthcare providers to focus on patient satisfaction and improve patient outcomes, rather than on diseases.

11 References

Abdalla, R., Pavlova, M., Hussein, M., & Groot, W. (2022). Quality measurement for cardiovascular diseases and cancer in hospital value-based healthcare: a systematic review of the literature. *BMC Health Services Research, 22*(1), 979. https://link.springer.com/article/10.1186/s12913-022-08347-x

Aljaloud, A., & Razzaq, A. (2023). Modernizing the legacy healthcare system to decentralize the platform using blockchain technology. *Technologies, 11*(4), 84. https://doi.org/10.3390/technologies11040084.

Alpert, J. S. (2023). Twenty-first-century healthcare challenges in the United States. *The American Journal of Medicine, 136*(7), 609-610. https://www.amjmed.com/article/S0002-9343(23)00035-9/fulltext

Alpert, J. S. (2023b). Remarkable advances in clinical medicine that have occurred since I was an intern. *The American Journal of Medicine, 136,* 499-500. https://www.amjmed.com/article/S0002-9343(23)00034-7/pdf

Akpegah, G., Ado, I., Agbama, J., & Essien, E. (2023). Patient-centered care: Current practice and recommendations for Nigeria. *Evidence-Based Health Policy, Management & Economics, 7*(3), 172-174. DOI:10.18502/jebhpme.v7i3.14280.

Alsalamah, S. (2021). Patient centered medical homes: Are they the right choice? *Saudi Journal of Nursing and Health Care, 4*(3), 84-85. DOI:10.36348/sjnhc.2021.v04i03.006.

Bailit, M., Hughes, C., & Bailit Health Purchasing, LLC. (2011). Key design elements of Shared-Savings payment Arrangements. In *Issue Brief: Vol. Vol. 20* [Issue Brief]. Bailit Health Purchasing, LLC. https://www.commonwealthfund.org/sites/default/files/documents/__media_files_publications_issue_brief_2011_aug_1539_bailit_key_d esign_elements_sharedsavings_ib_v2.pdf

Bao, C., & Bardhan, I. R. (2022). Performance of accountable care organizations: Health information technology and quality–efficiency

trade-offs. *Information Systems Research, 33*(2), 1-41.
http://dx.doi.org/10.1287/isre.2021.1080.

Bekbolatova, M., Mayer, J., Ong, C. W., & Toma, M. (2024).
Transformative potential of AI in healthcare: Definitions,
applications, and navigating the ethical landscape and public
perspectives. *Healthcare, 12*(2), 125.
https://doi.org/10.3390/healthcare12020125.

Belenguer, L. (2022). AI Bias: Exploring discriminatory algorithmic
decision-making models and the application of possible machine-
centric solutions adapted from the pharmaceutical industry. *AI and
Ethics, 2*(4), 771-787. https://doi.org/10.1007/s43681-022-00138-8.

Bendix, J. (2021). *The challenges, and opportunities, of clinically
integrated networks.* https://www.medicaleconomics.com/view/the-
challenges-and-opportunities-of-clinically-integrated-networks.

Cantor, M. N. (2020). Modernizing medical Attribution. *Journal of
General Internal Medicine, 35*, 3691-3693.
https://doi.org/10.1007/s11606-020-05838-7.

Centers for Medicare & Medicaid Services (CMS). (n.d). Quality
strategy goals. https://www.cms.gov/Medicare/Quality-Initiatives-
Patient-Assessment-
Instruments/QualityInitiativesGenInfo/Downloads/CMS-Quality-
Strategy-Goal-Card.pdf

Centers for Medicare & Medicaid Services (CMS). (2009). Medicare's
success in achieving major goals.
https://www.commonwealthfund.org/sites/default/files/documents/__
_media_files_publications_data_brief_2009_nov_1335_stremikis_hc
olmedicarereform_chartpack.pdf

Centers for Medicare & Medicaid Services (CMS). (2024a). Basics of
value-based care. https://www.cms.gov/priorities/innovation-
center/value-based-care-spotlight/basics-value-based-care

Centers for Medicare & Medicaid Services (CMS). (2024b). Health
equity. https://www.cms.gov/pillar/health-equity

Chen, Z. (2023). Ethics and discrimination in artificial intelligence-
enabled recruitment practices. *Humanities and Social Sciences*

Communications, 10(1), 1-12. https://doi.org/10.1057/s41599-023-02079-x.

de Levante Raphael, D. (2022). The knowledge and attitudes of primary care and the barriers to early detection and diagnosis of Alzheimer's disease. *Medicina, 58*(7), 906. https://www.mdpi.com/1648-9144/58/7/906

Dicpinigaitis, P. V., Altman, K. W., Ulger Isci, I., Ke, X., & Blaiss, M. (2023). Interdisciplinary collaboration in the diagnosis and management of chronic cough: The role and importance of primary care providers. *Current medical research and opinion, 39*(10), 1375-1381. https://doi.org/10.1080/03007995.2023.2255128

Dowd, B. E., & Laugesen, M. J. (2020). Fee-for-service payment is not the (main) problem. *Health Services Research, 55*(4), 491–495. https://doi.org/10.1111/1475-6773.13316.

Downey, S., Indulska, M., & Sadiq, S. (2019). Perceptions and challenges of EHR clinical data quality. *ACIS 2019 Proceedings. 25*. https://aisel.aisnet.org/acis2019/25.

Ghalavand, H., Shirshahi, S., Rahimi, A., Zarrinabadi, Z., & Amani, F. (2024). Common data quality elements for health information systems: a systematic review. *BMC Medical Informatics and Decision Making, 24*(1), 243. https://doi.org/10.1186/s12911-024-02644-7.

Gomez, T., Anaya, Y. B., Shih, K. J., & Tarn, D. M. (2021). A qualitative study of primary care physicians' experiences with telemedicine during COVID-19. *The Journal of the American Board of Family Medicine, 34*(Supplement), S61-S70. https://www.jabfm.org/content/jabfp/34/Supplement/S61.full.pdf

Gondi, S., Wright, A. A., Landrum, M. B., Meneades, L., Zubizarreta, J., Chernew, M. E., & Keating, N. L. (2021). Assessment of patient attribution to care from medical oncologists, surgeons, or radiation oncologists after newly diagnosed cancer. *JAMA Network Open, 4*(5), e218055-e218055. https://doi.org/10.1001/jamanetworkopen.2021.8055

Gupta, A. (2021). Impacts of Performance pay for hospitals: The Readmissions Reduction Program. *American Economic Review, 111*(4), 1241–1283. https://doi.org/10.1257/aer.20171825

Health Recovery Solutions (HRS). (2024). A quick guide to CMS value-based care models. https://www.healthrecoverysolutions.com/blog/a-quick-guide-to-cms-value-based-care-models

Kluwer, W. (2022, September 6). Six challenges to delivering quality healthcare. *Wolters Kluwer.* https://www.wolterskluwer.com/en/expert-insights/six-challenges-to-delivering-quality-healthcare

Kocakulah, M. C., Austill, D., & Henderson, E. (2021). Medicare cost reduction in the US: A case study of hospital readmissions and value-based purchasing. *International Journal of Healthcare Management, 14*(1), 203-218. https://doi.org/10.1080/20479700.2019.1637068

Kriznik, N. M., Lamé, G., & Dixon-Woods, M. (2019). Challenges in making standardization work in healthcare: lessons from a qualitative interview study of a line-labelling policy in a UK region. *BMJ open, 9*(11), e031771. https://doi.org/10.1136/bmjopen-2019-031771.

Leao, D. L. L., Cremers, H. P., van Veghel, D., Pavlova, M., & Groot, W. (2023). The impact of value-based payment models for networks of care and transmural care: A systematic literature review. *Applied Health Economics and Health Policy, 21*(3), 441-466. doi:10.1007/s40258-023-00790-z.

Lewis, V. A., McClurg, A. B., Smith, J., Fisher, E. S., & Bynum, J. P. (2013). Attributing patients to accountable care organizations: performance year approach aligns stakeholders' interests. *Health Affairs, 32*(3), 587-595. https://doi.org/10.1377/hlthaff.2012.0489.

Lopes, L., Montero, A., Presiado, M., & Hamel, L. (2024, March 1). Americans' challenges with healthcare costs. *Kaiser Family Foundation.* https://www.kff.org/health-costs/issue-brief/americans-challenges-with-health-care-costs/

Matenge, S., Sturgiss, E., Desborough, J., Hall Dykgraaf, S., Dut, G., & Kidd, M. (2022). Ensuring the continuation of routine primary care during the COVID-19 pandemic: a review of the international literature. *Family Practice, 39*(4), 747-761. https://doi.org/10.1093/fampra/cmab115

McCoy, R. G., Bunkers, K. S., Ramar, P., Meier, S. K., Benetti, L. L., Nesse, R. E., & Naessens, J. M. (2018). Patient attribution: why the method matters. *The American Journal of Managed Care, 24*(12), 596. https://pmc.ncbi.nlm.nih.gov/articles/PMC6549236/.

Moy, H.P., Giardino, A.P., & Varacallo, M.A. (2023). *Accountable care organization.* Treasure StatPearls.

Morenz, A. M., Zhou, L., Wong, E. S., & Liao, J. M. (2023). Association between capitated payments and preventive care among U.S. adults. *AJPM Focus, 2*(3), 100116. https://doi.org/10.1016/j.focus.2023.100116

NEJM Catalyst. (2018, March 1). *What is pay for performance in healthcare?* NEJM. Retrieved January 30, 2025, from https://catalyst.nejm.org/doi/full/10.1056/CAT.18.0245

Noël, F. (2022). Accelerating the pace of value-based transformation for more resilient and sustainable healthcare. *Future Healthcare Journal, 9*(3), 226-229. https://doi.org/10.7861/fhj.2022-0118

Peter G. Peterson Foundation. (2024, January 3). Why are Americans paying more for healthcare? https://www.pgpf.org/blog/2024/01/why-are-americans-paying-more-for-healthcare

Pfaar, O., Peters, A. T., Taillé, C., Teeling, T., Silver, J., Chan, R., & Hellings, P. W. (2025). Chronic rhinosinusitis with nasal polyps: Key considerations in the multidisciplinary team approach. *Clinical and Translational Allergy, 15*(1), e70010. https://doi.org/10.1002/clt2.70010

Rawal, P., Seyoum, S., & Fowler, E. (2024, June 4). Advancing health equity through value-based care: CMS Innovation Center update. *Health Affairs.* https://www.healthaffairs.org/content/forefront/advancing-health-equity-through-value-based-care-cms-innovation-center-update

Riley, W., Love, K., & Wilson, C. (2023, March). Patient attribution—A call for a system redesign. In *JAMA Health Forum* (Vol. 4, No. 3, pp. e225527-e225527). American Medical Association. https://doi.org/10.1001/jamahealthforum.2022.5527

Saghafian, S., Song, L., Newhouse, J., Landrum, M. B., & Hsu, J. (2023). The impact of vertical integration on physician behavior and healthcare delivery: Evidence from gastroenterology

practices. *Management Science, 69*(12), 7158-7179.
https://doi.org/10.1287/mnsc.2023.4886

Seshamani, M., Brooks-LaSure, C., & Weiss, R. (2025). A Stronger Medicare Program—Now and into the future [Dataset]. In *Forefront Group*. Health Affairs.
https://doi.org/10.1377/forefront.20250107.847453

Shih, T., Chen, L. M., & Nallamothu, B. K. (2015). Will bundled payments change health care? *Circulation, 131*(24), 2151–2158.
https://doi.org/10.1161/circulationaha.114.010393

Silva Almodóvar, A., Keller, M. S., Lee, J., Mehta, H. B., Manja, V., Nguyen, T. P. P., ... & Linsky, A. M. (2024). Deprescribing medications among patients with multiple prescribers: A socioecological model. *Journal of the American Geriatrics Society, 72*(3), 660-669.
https://agsjournals.onlinelibrary.wiley.com/doi/full/10.1111/jgs.18667

Sinsky, C. A., Bavafa, H., Roberts, R. G., & Beasley, J. W. (2021). Standardization vs customization: finding the right balance. *The Annals of Family Medicine, 19*(2), 171-177.
https://doi.org/10.1370/afm.2654

Society of Actuaries (2018). *Patient attribution the basis for all value-based care.* Retrieved March 20, 2025, from https://www.soa.org/globalassets/assets/Files/resources/research-report/2018/patient-attribution.pdf.

Tandan, M., Dunlea, S., Cullen, W., & Bury, G. (2024). Teamwork and its impact on chronic disease clinical outcomes in primary care: a systematic review and meta-analysis. *Public Health, 229*, 88-115.
https://doi.org/10.1016/j.puhe.2024.01.019

Teisberg, E., Wallace, S., & O'Hara, S. (2020). Defining and implementing value-based health care: a strategic framework. *Academic Medicine, 95*(5), 682-685.
https://doi.org/10.1097/ACM.0000000000003122.

Torvinen, H., & Jansson, K. (2023). Public health care innovation lab tackling the barriers of public sector innovation. *Public Management Review, 25*(8), 1-40. https://doi.org/10.1080/14719037.2022.2029107.

Teisberg, E., Wallace, S., & O'Hara, S. (2020). Defining and implementing value-based health care: A strategic framework. *Academic Medicine*, *95*(5), 682-685. doi:10.1097/ACM.0000000000003122

Wang, Z., Penning, M., & Zozus, M. (2019). Analysis of anesthesia screens for rule-based data quality assessment opportunities. *Studies in Health Technology and Informatics*, *257*, 473. https://europepmc.org/abstract/MED/30741242.

Wilson, M., Guta, A., Waddell, K., Lavis, J., Reid, R., & Evans, C. (2020). The impacts of accountable care organizations on patient experience, health outcomes, and costs: A rapid review. *Journal of Health Services Research & Policy*, *25*(2), 130-138. http://dx.doi.org/10.1177/1355819620913141.

Winberg, D. R., Baker, M. C., Hu, X., & Horvath, K. A. (2024). Who participates in value-based care models? Physician characteristics and implications for value-based care. *Health Affairs Scholar*, *2*(8), qxae087. https://doi.org/10.1093/haschl/qxae087

Zhang, Y., Callaghan-Koru, J. A., & Koru, G. (2024). The challenges and opportunities of continuous data quality improvement for healthcare administration data. *JAMIA open*, *7*(3), ooae058. https://doi.org/10.1093/jamiaopen/ooae058

www.ingramcontent.com/pod-product-compliance
Lightning Source LLC
Chambersburg PA
CBHW051444050726
47593CB00005B/1923